GEOECOLOGY

Animals, plants, and soils interact with one another. They also interact with the terrestrial spheres – the atmosphere, hydrosphere, toposphere, and lithosphere – and with the rest of the Cosmos. On land, this rich interaction creates landscape systems or geoecosystems.

Geoecology investigates the structure and function of geoecosystems. Part I introduces geoecological systems, their nature, hierarchical structure, and ideas about their interdependence and integrity. A simple dynamic systems model, referred to as the 'brash' equation, is developed to provide an analytical and conceptual framework for the book. Part II explores internal or 'ecological' interactions between geoecosystems and their near-surface environment, with individual chapters looking at the influence of climate, altitude, topography, insularity, and substrate. Part III prospects the role of external factors, both geological and cosmic, as agencies disturbing the dynamics of the geoecosystems.

A new 'evolutionary' view of geoecological systems, and the animals, plants and soils comprising them, emerges: geoecosystems are seen as dynamic entities, organized on a hierarchical basis, that perpetually respond to changes within themselves and in their surroundings.

Presenting a new ecological and evolutionary approach to the study of geoecological change, *Geoecology* will interest a wide range of environmental scientists, geographers, ecologists, and pedologists.

Richard John Huggett is a Senior Lecturer in Geography at the University of Manchester.

GEOECOLOGY

An evolutionary approach

Richard John Huggett

London and New York

First published 1995
by Routledge
11 New Fetter Lane, London EC4P 4EE

Simultaneously published in the USA and Canada
by Routledge
29 West 35th Street, New York, NY 10001

Typeset in Garamond by
Solidus (Bristol) Limited
Printed and bound in Great Britain by
Biddles Ltd, Guildford and King's Lynn

British Library Cataloguing in Publication Data
A catalogue record for this book is available from the British Library

Library of Congress Cataloguing in Publication Data
Huggett, Richard J.
Geoecology: an evolutionary approach/Richard John Huggett.
p.   cm.
Includes bibliographical references and index.
1. Biogeomorphology. 2. Ecology. I. Title.
QH542.5.H84   1995
574.5'22–dc20   94-30627

ISBN 0-415-08689-2
0-415-08710-4 (pbk)

For Jamie, Sarah, Edward, Daniel, Zoë, and Ben

CONTENTS

PLATES

PLATES

FIGURES

TABLES

PREFACE

This book, more than any other I have written, draws on my fascination with all aspects of the natural world. As a child I was interested in animals and plants, rocks and minerals, and maps. I spent many hours on wet Saturday afternoons with my cousin, now an exploration geologist, peering into the cases at the British Museum (Natural History), as it then was, and the Geological Museum, as well as pestering the staff at Gregory Botley's for small crystals and fossils. At secondary school, my interest in natural history took more formal shape and led to my taking geography, geology, zoology, and art at Advanced Level. I moved on to University College London where I read for a degree in geography, a subject that seemed wide enough in scope to embrace all my interests and more. The opportunity to specialize in physical geography courses presented itself and I took it eagerly. I still regard myself as a physical geographer, and do not admit to a narrower specialism than that. Research for my doctoral thesis, also carried out at University College, explored the idea of soil-landscape systems. To an extent, the present book is a belated development of that postgraduate work. An advantage of waiting so long to expand my original ideas on landscapes is that I have had time to mull over issues and read a lot. This means that I am clear in my own mind how interdependence in landscape systems might usefully be viewed and analysed.

The components of landscape systems are studied by scientists from disparate disciplines. In trying to give an integrated picture of landscape structure and dynamics, unity must be sought by offering an inter-disciplinary approach. My background in physical geography helps in doing this, though the reader will have to judge the success of my endeavours. Of course, the big problem with writing across traditional disciplinary bound-aries is in dishing out enough meat for the specialists. I have tried to do this in the text, but anyone hungry for more information on a particular theme should follow up the references provided. The book offers an approach to the study of landscape systems. My aim in writing the book is not to cater for the diverse tastes of specialists, but to convey to an audience of upper level students and academics a way of thinking about landscapes.

PREFACE

I am indebted to several people whose direct and indirect assistance in the production of this book is gratefully acknowledged: for drawing the diagrams, Graham Bowden (who had a little help from Nick Scarle); for taking the book on, Tristan Palmer at Routledge; for recognizing the value of 'armchair' research, Professor Peter Dicken; for exploring imaginary natural and supernatural worlds so engagingly, Douglas Adams, Brian Aldiss, Stephen Donaldson, Julian May, Terry Pratchett, J. R. R. Tolkien, and many others; for long, heady, and beery discussions on all manner of issues touching on life, the universe, and analogue computers, Derek Davenport; and for keeping me in touch with reality, my wife and children.

Richard Huggett
Poynton
August 1994

PROLOGUE

LAND'SCAPE, a View or Prospect of a Country so far as the Eye will
carry
SYST'EM, properly a regular, orderly Collection or Composition of
many Things together
(N. Bailey, *An Universal Etymological English Dictionary*, 1790)

Animals, plants, and soils interact with one another. They also interact with
the terrestrial spheres – the atmosphere, hydrosphere, toposphere, and
lithosphere – and with the rest of the Cosmos. On land, this rich interaction
creates landscape systems or geoecosystems. This book investigates the
structure and function of geoecosystems. It does so using a simple dynamic
systems model, christened the 'brash' equation, as a conceptual and analytical
tool. The model suggests an ecological and evolutionary approach for
studying geoecosystem dynamics: ecological because it allows for reciprocity
between all geoecospheric factors (everything is connected to everything
else) and evolutionary because it allows that conditions within and outside
geoecosystems are ever changing. Briefly, geoecosystems are seen as dynamic
entities whose components are richly interdependent, that are organized on
a hierarchical basis, and that perpetually respond to changes within themselves and in their surroundings.

The approach makes several assumptions about the structure and function
of geoecosystems, their components and their environment. For structure, it
is assumed that the geoecosphere and all its component spheres may be
viewed as a hierarchy of spatial systems. For function, it is assumed that all
components of geoecosystems are richly and often non-linearly interdependent; that geoecosystems behave holistically, in that their behaviour is not
predictable from the behaviour of their components; and that forcing factors
in the lithosphere and the rest of the cosmosphere are constantly changing.
Putting these structural and functional assumptions together leads to the
suggestion that geoecosystems are dynamic spatial entities that perpetually
respond to changes in their surroundings. Thus emerges an evolutionary, as
opposed to developmental, view of geoecological systems, and the animals,

plants, and soils comprising them: geoecosystems constantly evolve to accommodate changes in their internal, cosmic, and geological environments. This evolutionary view provides a new way of thinking about and studying geoecological change.

The conceptual framework, around which the book is structured, is given by the 'brash' equation. In brief, the 'brash' equation is a set of equations describing the dynamics of the geoecosphere. The geoecosphere is defined as interacting terrestrial life and life-support systems – the biosphere, b, toposphere, r, atmosphere, a, pedosphere, s, and hydrosphere, h. The time rate of change of each geoecospheric component depends on the state of all others, plus the effect of cosmic, geological, and other forcing factors, z, which lie outside the geoecosphere. When expressed mathematically, these ideas yield the 'brash' equation:

$$\frac{db}{dt} = f(b, r, a, s, h) + z$$

$$\frac{dr}{dt} = f(b, r, a, s, h) + z$$

$$\frac{da}{dt} = f(b, r, a, s, h) + z$$

$$\frac{ds}{dt} = f(b, r, a, s, h) + z$$

$$\frac{dh}{dt} = f(b, r, a, s, h) + z$$

This set of equations is a very general dynamical model of geoecosystems. It is a logical consequence of assuming that geoecosystems evolve from interactions within and between the terrestrial biosphere, toposphere, atmosphere, pedosphere, and hydrosphere. The 'brash' formula seems to handle interdependence of geoecosystem components more satisfactorily than Hans Jenny's classic 'clorpt' equation (in which a geoecosystem or geoecosystem property is defined as a function of climate, organisms, relief, parent material, and time), because reciprocity between all geoecosystem factors is assumed; and, unlike the 'clorpt' equation, it expressly treats time as a truly independent variable that affects all factors. The advantage of the 'brash' formula is that it supplies an analytical, as well as a conceptual, framework for studying geoecospheric change: it represents geoecosystem structure, function, and dynamics in a mathematical form amenable to dynamic systems analysis, and to analysis by less rigorous, but often very revealing, multivariate statistical techniques.

The above ideas are developed and illustrated in the book. The discourse is tripartite. Part I introduces geoecosystems, describing their nature, hierarchical structure, and ideas about their interdependence and integrity. In addition, it develops the 'brash' equation, the model that provides the conceptual framework for the book. The rest of the book is concerned with internal and external influences on life and soils within geoecosystems. Part II is the core of the book. It explores internal or 'ecological' interactions between geoecosystems and their near-surface environment. Individual chapters deal with the environmental factors listed in the 'brash' equation: climate (atmosphere and hydrosphere), topography (toposphere), and substrate (pedosphere and lithosphere). Chapters 3 and 4 consider latitudinal and longitudinal, and Chapter 5 altitudinal, climatic components of geoecosystems. Chapter 6 looks at substrate as a component of geoecosystems. Chapters 7 and 8 examine the topographic component of geoecosystems – Chapter 7 probing the effect of aspect, slope gradient, slope curvature, and contour curvature on animals, plants, and soils, and Chapter 8 investigating the effect of insularity, which is basically a topospheric property, on animals, plants, and communities. Part III prospects the role of external factors (ecological, geological, and cosmic) as agencies disturbing the dynamics of geoecosystems.

The discerning reader will possibly have noticed that this discursive framework follows a well-trodden path. It is similar to the framework used by Hans Jenny (1980) when discussing his 'clorpt' equation. The present book makes an impassioned plea for the adoption of a model of geoecosystems that is at once ecological and evolutionary. Given this professed aim, it may seem somewhat craven and self-defeating to look at environmental influences on life and soils in geoecosystems by singling out different environmental factors and considering them one by one. The reasons for doing this are simple. First, much existing work looks at the effects of individual factors. Second, it is convenient to organize the material on a factor-by-factor basis. Third, studies that take on board a range of environmental factors and launch a multivariate attack on the data can be usefully discussed within a univariate framework. Indeed, they often show that just a few particular factors do seem to wield a major influence at a particular scale, albeit in a more cryptic way than was hitherto understood. The framework does not ignore the multivariate interdependence expressed in the 'brash' equation; it simply recognizes that, for a particular geoecosystem, the ecological relationships and evolutionary changes are strongly influenced by a particular group of variables, either internal or external. None the less, as will become apparent, a deep appreciation of geoecological dynamics does require a knowledge of the whole environmental complex and the rich web of interdependencies contained therein.

Part I

INTRODUCING GEOECOSYSTEMS

1

TERRESTRIAL SPHERES

In recognizing four basic terrestrial elements – air, water, earth, and fire – the ancient Greek philosophers identified, without naming them, the chief spheres of the Earth: the gaseous sphere, or atmosphere; the watery sphere, or hydrosphere; and the solid sphere, or lithosphere. Fire, the fourth element, has no modern counterpart, but it was originally conceived, not as a zone around the atmosphere that burns brightly, but as a region where fire has a propensity to break out. The sky appears to burn when lightning flashes and when meteors enter the atmosphere and explode as fireballs. It is perhaps significant that the activity of meteors is tame today compared with times in human history when the brilliant illumination of the night sky was a regular occurrence (Clube and Napier 1990).

Before 1875, the only sphere given a special name was the atmosphere. Then the Austrian geologist Eduard Suess, in the last and most general chapter of a slim volume entitled *Die Entstehung der Alpen* (The Origin of the Alps), invented the eminently helpful terms hydrosphere, lithosphere, and biosphere. Since then, Earth and life scientists have gone somewhat 'sphere crazy', and many parts of the Earth and its cosmic environment are given labels suffixed with the term 'sphere'. Examples include cosmosphere, pedosphere, ecosphere, landscape sphere, rhizosphere, barysphere, centrosphere, and bathysphere. The list is large. Humans possess a fondness for recognizing and naming objects in Nature. Perhaps the word 'sphere' has proved so serviceable because, in combination with suitable prefixes, it provides memorable and punchy terms for parts of the Earth.

Today, interest focuses on the interactions between the terrestrial spheres. One way of examining these interactions, and the systems that they produce, is to follow the lead given by Sante Mattson (1938) who considered all possible interactions between the lithosphere, atmosphere, hydrosphere, and biosphere (Figure 1.1). A different schema of cosmic and terrestrial spheres and their interaction is suggested in Figure 1.2. The cosmosphere is the domain of all non-living things and forces and includes the Earth. The Earth is closely associated with objects in the rest of the Cosmos in at least three ways: it is a recipient of energy generated by stars; it is a component in the

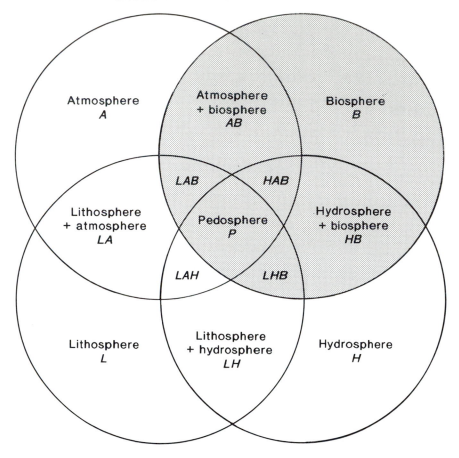

Figure 1.1 Terrestrial spheres and their interaction as envisioned by Sante Mattson. The shaded portion is the ecosphere, a term unknown to Mattson. Examples of the interacting zones suggested by Mattson are: *LA*, a barren desert; *AB*, the aerial space between plants; *HB*, a pond; *LH*, waterlogged sand or clay under sterile conditions; *LAB*, guano deposits; *HAB*, organic soils and forest litter; *LHB*, waterlogged soils and lake bottoms; *LAH*, very saline soils

Source: After Mattson (1938)

gravitational field of the Solar System, Galaxy, and Universe; and it is a potential target for space debris. The Earth itself consists of several terrestrial spheres. As the terms associated with these spheres are, in some cases, ambiguous, their usage will be examined before proceeding to study their interaction in geoecosystems.

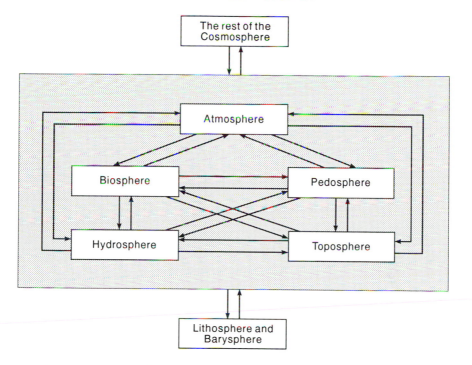

Figure 1.2 A schema for the terrestrial spheres, their interaction and external influences

GEOSPHERES

The term 'geosphere' has three meanings (Bates and Jackson 1980). First, it is simply the lithosphere. Second, it is the lithosphere, hydrosphere, and atmosphere combined. And third, it is any of the terrestrial spheres or shells. It is difficult to gauge which of these meanings is the most commonly used. Herbert Friedman (1985) offers a fuller, and therefore wordier, definition: the geosphere is the totality of geophysical systems comprising the lithosphere, hydrosphere, troposphere, stratosphere, mesosphere, thermosphere, exosphere, ionosphere, and magnetosphere. That seems to cover all abiotic spheres, except that it does not expressly include the solid Earth below the lithosphere (depending on how the lithosphere is defined). In this book, the geosphere is taken to include the core, mantle, and all layers of the crust.

Lithosphere

Since it was first used in 1875 to describe the solid Earth, the term 'lithosphere' has acquired two meanings, both of which are useful. In a general sense, the lithosphere is the solid portion of the Earth – the rocks. Many geologists writing before the advent of plate tectonics adhered to this meaning. Thus, in *Lake and Rastall's Textbook of Geology*, it states that:

> From the geological point of view the earth may be regarded as consisting of two concentric shells and a central sphere, of very different natures. As a matter of convenience the two shells may also be called spheres, though that is not strictly correct: the three components are then the atmosphere, the hydrosphere and the lithosphere ... The third of these spheres, the lithosphere, is the solid earth, and it is essentially the province of geology to study its structures and history.
>
> (Rastall 1941: 2)

On the other hand, some writers elected to limit the lithosphere to the outer shell of the solid Earth, where the rocks are more or less similar to those exposed at the surface. The inner portion of the solid Earth was thus distinguished from the lithosphere and variously styled the centrosphere, barysphere, and bathysphere (easily confused with a submersible used to explore the ocean depths), or even pyrosphere and magmosphere. The barysphere may refer to the mantle, or to the core, or to both. Since the coming of plate tectonics, the practice of defining the lithosphere more narrowly to mean the relatively strong surficial layer of the solid Earth lying above the relatively weaker asthenosphere is commonplace. The lithosphere of plate tectonics includes the crust and the solid part of the upper mantle and is, on the average, about 100 km thick. Below continents it is some 150 km thick, and beneath the oceans it is some 60 km thick. Increased knowledge of the Earth's interior has led to the waning of catch-all terms such as barysphere. It is more normal to use the names given to the chief divisions of the solid Earth as revealed by seismic data. Thus, the lithosphere sits atop a 'weak' layer, or asthenosphere, that is part of the upper mantle. The rocks within it may be partially molten and, over protracted time periods, act as fluids, so allowing the lithospheric plates to glide serenely over the Earth's surface. Between 400 and 650 km below the surface, rocks again become harder in the transition zone to the lower mantle. Extending down to a depth of 2,890 km, the lower mantle accounts for nearly one-half of the Earth's mass. It lies upon the Earth's core, into which it merges through a fairly sharp transition zone known as the D″ layer. The core consists of an outer molten shell, some 2,260 km thick, and a solid inner ball 1,228 km in radius. Processes occurring in the core and mantle influence plate tectonics, and so, indirectly, they may eventually cause changes in geoecosystems.

Atmosphere

The word atmosphere was first used in 1638 to describe an orb of vapour that was supposed to enshroud the Moon. It was soon applied to the ring of 'vaporous air' presumed to be exhaled from the body of a planet. This 'vaporous air' was deemed to be part of the planet whereas the surrounding air was not. By the end of the seventeenth century, the atmosphere had come to mean all air within a planet's sphere of activity. This meaning survives today. The atmosphere is the shell of aeriform fluid that envelops the Earth. It is a dusty gas, much of the mass of which is contained in its lowermost layer. Customarily, it is divided into several spheres, each of which has a characteristic temperature, pressure, and composition: the troposphere, stratosphere, mesosphere, thermosphere, and exosphere. Besides these divisions, there is the ionosphere, a shell of high electron concentration; and, extending well out into space, is the constantly changing magnetic field generated by the Earth's dynamo, what Thomas Gold dubbed the magnetosphere. The weather that affects geoecosystems is confined to the relatively dense troposphere. Several complex chemical and thermal reactions take place in the rare upper air. They are powered by the incoming flux of waves and particles from the cosmosphere. These reactions seem to have a far greater influence on the climate at the ground than was once thought possible, though the connections between the two are still a trifle puzzling.

Hydrosphere

The hydrosphere is the entirety of the waters of the Earth. It includes liquid water, water vapour, ice and snow. Waters in the oceans, in rivers, in lakes and ponds, in ice sheets, glaciers, and snow fields, in the saturated and unsaturated zones below ground, and in the air above ground are all part of the hydrosphere. Some people set the ambits of the hydrosphere to exclude the waters of the atmosphere. The hydrosphere presently holds about 1,384,120,000 km^3 of water in various states, most of which is stored in the oceans. A mere 2.6 per cent (36,020,000 km^3) of the hydrosphere is fresh water. Of this, 77.23 per cent is frozen in ice caps, icebergs, and glaciers. Groundwater found above a depth of 4 km accounts for another 22.21 per cent of fresh water. The tiny remainder is stored in the soil, lakes, rivers, the biosphere, and the atmosphere.

Toposphere

The German geomorphologist Julius Büdel (1982) invented the term 'relief sphere' to describe the totality of the Earth's topography. The term 'toposphere' is proposed here as a more euphonic substitute. The toposphere sits at the interfaces of the pedosphere and atmosphere and pedosphere and hydrosphere.

As some confusion surrounds the use of the terms 'relief' and 'topography', a token explanation seems in order. In lay and professional circles, it is common to use the words 'relief' and 'topography' commutably, but this practice is to be frowned upon in technical writing since it can lead to misconstruals. Topography is not a problem word because, although it is used in more than one way, ambiguity seldom arises. It means the lie of the land, or the general configuration of the land surface, including its relief and the location of its features, natural and man-made. It also expresses the physical surface features of a region as displayed by the contours on a map. Difficulties arise when using the word 'relief', primarily because it is sometimes brought into service as a synonym of topography. In a more restricted sense, relief is the vertical difference between the highest and lowest elevations in a region. To avoid confusion, it is perhaps better to say 'topographic relief' where topography is meant, and to restrict relief to elevational differences. No hard-and-fast recommendations are made here about how the words should be used. The reader is simply alerted to a potential area of confusion. But the confusion is not the idle fancy of the author, who openly admits being irresistibly drawn into semantic black holes. One has only to read some of the papers addressing the relief and topographic factor of pedogenesis to appreciate the difficulty.

BIOSPHERE AND ECOSPHERE

The Earth, so far as is known, is unique among the planets and satellites in the Solar System: it alone houses life; it alone boasts a biosphere. Life on Earth inhabits the lower parts of the air, the oceans, seas, lakes, and rivers, the land surface, and the soil. Life depends on its environment to survive: all life is dependent on mineral resources stored in the geosphere, and most of it upon sunlight. Equally, it is influenced by, and has to adapt to, other factors in its surroundings. It responds to forces and events originating in the Solar System and Galaxy. Likewise, it responds to forces and events springing from the Earth's interior. Through its interaction with its surroundings, life on Earth creates and conserves an ecosphere, a zone fit for terrestrial-type life-forms.

Unluckily, there is considerable confusion surrounding the meaning of the terms 'biosphere' and 'ecosphere'. The word 'biosphere' has three meanings: the totality of living things dwelling on the Earth, the space occupied by living things, or life and life-support systems – atmosphere, hydrosphere, lithosphere, and pedosphere (Table 1.1). If the biosphere is restricted to the totality of all living things, then another word is needed to describe all life and the inorganic environment that sustains it. LaMont C. Cole (1958) coined the term 'ecosphere' for that purpose. He apologized for using a coined word like ecosphere, but it seemed to him nicely to describe just what he wanted to discuss. His intention was to combine two concepts: the biosphere and the ecosystem. The biosphere he took to mean the totality of living creatures on the Earth. The ecosystem he took as a self-sustaining community of organisms (animals and plants) together with their inorganic

8

Table 1.1 Three meanings of the term 'biosphere'[a]

	Vital skin of the Earth	Integrated life and life-support entity	Space in which life resides
Introducer	Pierre Teilhard de Chardin	Vladimir Ivanovich Vernadsky	George Evelyn Hutchinson
Definition	The actual layer of vitalized substance enveloping the Earth; the totality of living beings	The unit, partly created and controlled by life, resulting from the coevolution of living things and their planetary environment[b]	That part of the Earth in which life exists
Source	Teilhard de Chardin (1969: 163; 1957, 1959)[c]	Vernadsky (1926, 1929, 1944, 1945)	Hutchinson (1965, 1970)
Supporters	Mill (1899: 4) Thomson (1931: 204, 205) Mattson (1938) Gillard (1969) Golley (1978)	Florkin (1943) Allee et al. (1949) Duvigneaud (1974) Dasmann (1976: 6) Grinevald (1988) Botkin (1990: 229)	Dansereau (1957: 125) Tivy (1982: 14) Bolin (1980: 3) de Blij and Muller (1993: 14)

Notes: [a]The term biosphere was invented by Eduard Suess who wrote of a sphere of living organisms or biological processes – 'eine selbständige Biosphäre' (an independent biosphere) – lying at the interface between the atmosphere, lithosphere, and hydrosphere (Suess 1875: 159)
[b]Vernadsky's notion of the biosphere is similar to notions of the biogeocoenose (e.g. Sukachev and Dylis 1968) and the ecosystem (Tansley 1935)
[c]Teilhard first used the term in a panegyrical review, published in 1921, of Suess's The Face of the Earth; he discussed it in essays written during 1925–1926 (see Teilhard de Chardin 1957)

environment. This notion was clearly inspired by Arthur George Tansley's (1935) image of an ecosystem: a self-sustaining community of organisms together with their physical environment. To Cole, the ecosphere is the global ecosystem, 'the sum total of life on earth together with the global environment and the earth's total resources' (L. C. Cole 1958: 84). Cole's term was later reinvented for describing 'that part of our sphere in which there is life together with the living organisms it contains' (Gillard 1969).

The term 'ecosphere' has been used by ecologists and biogeographers. Joy Tivy employed it in *Biogeography: A Study of Plants in the Ecosphere* (1982). Barry Commoner, in *The Closing Circle* (1972), used the idea of the ecosphere as a framework in which to consider the 'environmental crisis'. However, he also spoke of the ecosphere as 'the home that life has built for itself on the planet's outer surface' (1972: 11), a definition redolent of Hutchinson's biosphere (Table 1.1). It is evident that the term 'biosphere', as originally used by Vladimir Ivanovich Vernadsky (Table 1.1), is equivalent to the term 'ecosphere' as coined by Cole (and independently by Gillard). The obvious conclusion is that Cole and Gillard's ecosphere is redundant. However, the word 'ecosphere' seems to capture Vernadsky's conception of life and life-support systems better than does the word 'biosphere'. But let us not draw too hasty a conclusion: the ecosphere has an older claim to fame.

In 1953, Hubertus Strughold wrote a book called *The Green and Red Planet: A Physiological Study of the Possibility of Life on Mars*. In this book, he used the term 'ecosphere' to define the zones in the universe that would be habitable by living organisms:

> Only a small zone about 75 million miles wide – out of the 4,300 million that stretch between the sun and Pluto at its farthest point – provides a planetary environment well-suited to the existence of life. We might call this zone the thermal ecosphere of the sun.
>
> Other stars may have such ecospheres of their own, with planets in them that are capable of supporting life similar to ours.
>
> (Strughold 1953: 43)

Astronomers have subsequently used the word 'ecosphere' to mean regions in space where conditions would allow living things to exist (at least, living things as we know them). And it is Strughold's idea of an ecosphere, not Cole's, that is found in dictionaries. In the 1972 *Supplement to the Oxford English Dictionary* it is defined as 'The region of space including planets whose conditions are not incompatible with the existence of living things'. In the *Glossary of Geology*, it is described as 'Portions of the universe favorable for the existence of living organisms' (Bates and Jackson 1980). This definition has much to commend it – and it would not invalidate Cole's definition.

Plainly, the terms 'biosphere' and 'ecosphere' have had chequered histories. Literally, biosphere is a combination of βιοσ (life) and σπηαερα

10

(sphere). Vernadsky chose to define it as the functional sphere comprising all life and life-support systems. Teilhard de Chardin preferred to restrict it to the sphere of all living things. Hutchinson and others opted for the actual matter and space occupied and affected by life, now or in the past. There is no 'correct' definition. People use words to serve their needs in communicating ideas. It is, perhaps, unrealistic to expect words such as biosphere and ecosphere, with colourful pasts, ever to acquire standard meanings. Nicholas Polunin and Jacques Grinevald's (1988) valiant and sincere attempt to compose a universally agreeable definition of The Biosphere is to be much admired. (They bestow on the word a capital B to dignify our only known natural habitat in the Cosmos.) To them, The Biosphere is the 'integrated living and life-supporting system comprising the peripheral envelope of Planet Earth together with its surrounding atmosphere so far down, and up, as any form of life exists naturally' (Polunin and Grinevald 1988: 118). Vernadsky would doubtless have approved of this definition; Teilhard de Chardin might have had qualms about it. Who is right? Everybody is and nobody is. The biosphere and the ecosphere are productions of the human mind. Many different minds have deliberated upon their nature, hence they have been conceived of in various ways. For this reason, it is dangerous to be too prescriptive about definitions and pass judgement on their virtues. The important thing, surely, is not doggedly to follow a single definition, but to be aware of the variety of meanings conveyed by the words and to use them judiciously. That having been said, a position must be taken to avoid confusion in this book. In the following pages, the biosphere will mean the totality of living things, and the ecosphere will be a neutral term for the global sum of life and life-support systems. (Gaia is an unneutral term equivalent to ecosphere and, despite its creators' disclaimers, laden with vitalistic undertones.)

PEDOSPHERE

The pedosphere is 'that shell or layer of the Earth in which soil-forming processes occur' (Bates and Jackson 1980). But what are soil-forming processes? And what is soil? These vexatious questions have plagued pedologists since the foundation of their discipline. The issue is too important to be glossed over: an opinion must be passed. At the risk of causing ructions in upper echelons of Cambridge University, one might suggest that much can be learnt about a person's predilections towards soil and its relationships with other terrestrial spheres by 'deconstructing' his or her definition of it. Two conflicting definitions exist, one adopted by geologists and engineers, the other espoused and zealously guarded by pedologists. Geologists and engineers see soils as soft, unconsolidated rocks. According to this definition, the entire profile of weathered rock and unconsolidated rock material, of whatever origin, is soil material. Used in this way, soil is the same as regolith. Regolith means 'stony blanket' and fittingly describes many mantles of weathered materials. However, it has a wider compass than stony mantles and includes stone-free materials as well as organic materials such as peat.

Most pedologists see soils as that part of the regolith which supports plant life, and which is affected by soil-forming processes. The corollary of this, by a rather circular argument, is that soil-forming processes operate in that part of the regolith influenced by plant life – that is, the soil. There are several difficulties with this definition, as most pedologists themselves acknowledge. Some saline soils and laterite surfaces cannot support plants – are they then soils? Is a bare rock surface encrusted with lichens a soil? Pedologists cannot agree on these troublesome matters. Hans Jenny (1980: 364) owns that the lack of agreement is embarrassing, but he finds cheer in the fact that biologists cannot agree on a definition of life, nor philosophers on a definition of philosophy! He struggles to side-step the conceptual dilemma of what is rock and what is soil by suggesting that exposed hard rocks are soils (Jenny 1980: 47). The basis of this suggestion is that exposed rocks, like soils, are influenced by climate; and that, like some soils, they will support little or no plant life. To be sure, it seems impossible to pass through the horns of the dilemma: either soils are seen as rocks, or else rocks are regarded as soils. However, there is a case for falling in line with geologists and with Jenny and referring to all weathered material as soil. This avoids the semantic confusion of taking the opposite view and referring to soil as rock. If this latter course were followed, one would be faced with ambiguous terms. What, for instance, would be meant by rock-forming processes? No, it seems preferable to distinguish between rock in the lithosphere (its environment of formation), and rock exposed to ecospheric processes. There would seem to be merit in plumping for an interdisciplinary definition of soil. Something along these lines might fit the bill: soil is rock that has encountered the ecosphere. This definition of soil has a big virtue: it spotlights the unitary nature of processes in soil landscapes, and in doing so it calls into question the somewhat arbitrary distinctions between soil and regolith, and between soil processes and geomorphological processes. The unity of soils and landscapes finds expression in the notion of 'pedomorphic surfaces' (Dan and Yaalon 1968) and Bruce E. Butler's (1982) idea of the soil mantle and its basic unit, the pedoderm (Brewer et al. 1970). It also is in line with the notion of three-dimensional soil mantles expressed in the new French Référentiel Pédologique (e.g. Baize 1993).

Pedologists may dislike this definition of soil. They may think that it takes them a horizon too far. If so, then they can fall back on a term that already exists in their own vocabulary – solum. The solum is the genetic soil developed by soil-building forces (Soil Survey Staff 1975), and normally comprises the A and B horizons of a soil profile, that is, the soil and subsoil. It is the bit of the weathered mantle that most pedologists are interested in.

Pedologists may also take comfort in the fact that the definition of the soil profile would remain unadulterated; it would still be the pedogenetically altered material as well as the deep layers (the substrata) that have influenced pedogenesis. Given that the soil of the pedologist is a widely used and valuable concept, it seems sensible to regard it as a subsphere of the pedosphere and give it label. A germane expression is edaphosphere. The remaining portion of the pedosphere – all the material lying on the lithosphere but not including that encompassed by the edaphosphere – may be christened the debrisphere. The debrisphere is roughly equivalent to the decomposition sphere as designated by Julius Büdel (1982), but includes detritus created by mechanical disintegration, as well as the productions of chemical weathering (cf. Nikiforoff 1935). A lifeless planet may have a debrisphere because the surficial rocks of any terrestrial planet or satellite will be exposed to an atmosphere and that will lead to decomposition, disintegration, and the formation of a weathered mantle. This material is referred to by planetologists as soil, as in lunar soil and Martian soil. The planetologists' conception of soil agrees with the conception proposed here, but is at odds with the pedologists' conception. Planets and satellites devoid of life cannot, by definition, have an edaphosphere.

As the lithosphere has been excluded from the ecosphere, the status of 'parent material' needs clarification – is it part of the ecosphere or part of the geological environment? Before coming to a decision on this matter, it is worth recalling that organic materials produced by the biosphere are the parent material for some soils and plants. For this reason, there may be some merit in excluding the lithosphere as a whole from the ecosphere, but including parent material. Viewed in this way, parent material may be thought of as material from the lithosphere or biosphere lying within the influence of the ecosphere and being subject to alteration by it. This is roughly the debrisphere as defined earlier. Parent materials derived from the lithosphere exist in an unaltered state only in those parts of the lithosphere that the biosphere cannot reach; such material can be thought of as grandparent material. However, Karsten Pedersen (1993) has drawn together shreds of independent evidence showing that microbial life is widespread at depth in the lithosphere, where it may be involved in subterranean geochemical processes. In consequence, grandparent material may lie much deeper than is commonly supposed, possibly at depths exceeding about 4,200 m! The digging of a soil pit that deep would defy the most energetic soil surveyor.

GEOECOSPHERE

Current research into interdependence within the ecosphere focuses on the role of geographical space – the landscape. The word 'landscape' commonly alludes to many different things: a picture of a view of natural inland scenery,

Table 1.2 Scales and terminology of landscape systems

Scale[a]	Approximate area (km²)	Terminology applied to these scales[b]		
		Fenneman (1916)	Linton (1949)	Whittlesey (1954)
Micro (small)	< 10⁰	–	Site	–
Meso (medium)	10⁰–10¹	–	–	–
	10¹–10²	–	Stow	Locality
	10²–10³	District	Tract	District
	10³–10⁴	Section	Section	–
Macro (large)	10⁴–10⁵	Province	Province	Province
	10⁵–10⁶	Major division	Major division	Realm
Mega (very large)	>10⁶	–	Continent	–

Notes: [a]These divisions follow Delcourt and Delcourt (1988)
[b]The range of areas associated with these regional landscape units are meant as a rough-and-ready guide rather than precise limits

a vista of natural scenery seen by the eye, and the landforms of a region seen as a whole. From an ecological perspective, the landscape is the land surface and its associated habitats viewed at medium scales (Table 1.2); or, simply, a spatially heterogeneous area or environmental mosaic (M. G. Turner and Gardner 1991). In an ecological context, the landscape may fruitfully be viewed as a sphere within which the other terrestrial spheres interact – the landscape sphere (Vink 1983: 1). Viewed as a system, the landscape sphere (or geoecosphere) is the dynamic product of interacting ecospheric systems. And, a landscape (or geoecological) system may be defined as any landscape unit in which the biosphere, toposphere, atmosphere, pedosphere, and hydrosphere, together with the biological, geomorphological, climatological, pedological, and hydrological processes that create them, are seen as a unitary whole. All geoecosystems have structure and function and are dynamic (cf. Forman and Godron 1986: 11).

Geoecosystems

Present interest in landscapes springs from regional geography and vegetation science as practised by European workers, and from the fascination of many Russian geographers, geologists, and pedologists with processes in landscapes. The biggest strides in advancing the knowledge of geoecological processes over the last two decades have been taken by landscape ecologists. The term 'landscape ecology' was devised by Carl Troll (1939) to marry geography (the landscape) with ecology. Later, he coined the word 'geoecology' to describe the same field of study (Troll 1939, 1971, 1972). Although mainland Europeans have a long tradition of landscape ecology, North

Americans have rediscovered it only in the last decade.

Modern landscape ecologists probe the causes and effects of spatial patterning in ecosystems. Specifically, they consider four aspects of landscape systems (M. G. Turner 1989). First, they investigate the evolution and dynamics of spatial heterogeneity – how the landscape mosaic is created and how it changes. Second, they look at interactions between, and exchanges across, heterogeneous landscapes – how materials and organisms move from one patch to another. Third, they elucidate the influence that the spatial heterogeneity of the landscape mosaic has upon biotic and abiotic processes in the landscape. And fourth, they consider the management of spatial heterogeneity. The older, Clementsian ecological paradigm was non-spatial. In the early decades of the twentieth century, many schools of phytosociology followed Frederic E. Clements's (e.g. 1916, 1936) lead in recognizing homogeneous or uniform communities and displaying disinterest in spatial change (McIntosh 1991: 30). In stark contrast, Herbert A. Gleason, the creator of the so-called individualistic concept, evinced a strong interest in spatial patterns (e.g. Gleason 1926). He argued that each species has its individual requirements, that the environment varies continuously, and that, putting these two assumptions together, the community is an individualistic admixture of species and environment thrown together by happenstance. These ideas were 'heterogeneity rampant' and were studiously eschewed by phytosociologists until the 1950s (McIntosh 1991: 32). A turning point in the acceptance of landscape heterogeneity seems to have been a paper by Alexander Stuart Watt published in 1947, although in 1924 he had described the beechwood communities on the Sussex Downs as a mosaic of patches of different ages resulting from small-scale disturbances at different intervals. Watt (1947) outlined several examples in which apparently homogeneous (climax) communities undergo continual and cyclical change involving pioneer, building, mature, and degenerate phases. In essence, he saw that the steady-state pattern in the landscape as a whole was maintained by a constantly changing state in individual landscape patches: the landscape is heterogeneous.

Building on Watt's framework, landscape ecologists now deal with processes occurring at a wide range of spatial and temporal scales. Their viewpoint is largely biological: they tend to focus on the biotic component of geoecosystems, though there are many exceptions to this generalization. Interest in the abiotic portion of landscapes has its roots in physical geography, geomorphology, geology, and pedology. The Russian geochemist, Boris B. Polynov, believed in the integrity of the landscape in producing, transporting, and removing rock debris. His chief concern was with the interaction of the terrestrial spheres, which, he deemed, determines the migration of chemical elements in a landscape (e.g. Polynov 1935, 1937). His ideas spawned a Russian school of landscape geochemists who focused their attention on the flow of matter though landscapes. Eventually, the ideas of the Russian school went west and several geochemists became interested in chemical migration through landscapes (e.g. Rose *et al.* 1979; Fortescue 1980). Independently of these developments in landscape geochemistry, the concept of the soil-landscape system was served up (Huggett 1973, 1975). A soil-landscape system is 'any landscape unit in which landforms and soils, and the geomorphological and pedological processes which create them, are seen as a unitary whole' (Huggett 1991: 278). This concept was designed to link soil processes and geomorphological processes in a landscape, a theme pursued by pedologists with a geomorphological leaning, and geomorphologists with a keen interest in soils (e.g. Ruhe and Walker 1968; Conacher and Dalrymple 1977; Gerrard 1981, 1992, 1993).

The concept of geoecosystems, as set down in the present book, widens the concept of soil-landscape systems to embody animals and plants. It underscores the connection between the biotic and abiotic components of a geoecosystem. In doing so, it serves to counterbalance the biological emphasis on landscapes evinced by many landscape ecologists. Having said that, a geoecosystem may be defined in the same way that landscape ecologists define landscape – that is, as 'a heterogeneous land area composed of a cluster of interacting ecosystems that is repeated in similar form throughout' (Forman and Godron 1986: 11). By way of example, the recurring cluster of interacting ecosystems that feature in the landscape around the author's home, in the foothills of the Pennines, includes woodland, field, hedgerow, pond, brook, canal, road, path, disused mining incline, and disused railway. The elements, or fundamental units, comprising a geoecosystem are variously termed ecotopes, biotopes, geotopes, facies, habitats, sites, tesserae, landscape units, landscape cells, landscape prisms, or simply landscape elements. References to most of these may be found in Richard Forman and Michel Godron's book, *Landscape Ecology* (1986). The landscape prism was designed by John A. C. Fortescue (1980: 12) to integrate relations between geology, soil, and vegetation at a particular locale. It is a small spatial unit of limited horizontal extent with vertical sides centred on the pedosphere, extending downwards to the lithosphere and upwards through the biosphere to the atmosphere. Of all the proposed terms, tesserae is the most appealing since, like the basic pieces of stone in a decorative mosaic, tesserae are homogeneous components of landscape mosaics (cf. Jenny 1958, 1965); but perhaps it is preferable to refer to them by the bland term landscape elements. Whatever they be called, landscape elements are 'the smallest homogeneous landscape unit[s] visible at the spatial scale of a landscape' (Forman and Godron 1986: 13).

It is important to note that landscapes of landscape ecologists are normally limited to a fairly narrow range of spatial scales: landscape elements are no smaller than about 10 m and whole landscapes are no bigger than about 10,000 km². In contrast, the concept of geoecosystems developed in the present book applies to all levels in the hierarchy of geoecosystems. Its compass ranges from tesserae a square metre or less in area, through regions and continents, to the entire geoecosphere. It is worth noting that some landscape ecologists appear to be relaxing their interpretation of a landscape to include smaller and larger scales (T. F. H. Allen and Hoekstra 1992: 55). They have come to realize that an ant's view and a bird's view of the landscape is very different to a human's view. Happily, many of the principles and ideas of landscape ecology can be translated to landscapes at smaller and larger scales. Landscape ecologists have taken pains to thrash out a common vocabulary and a set of working definitions of terms pertaining to scale (M. G. Turner *et al.* 1989). As this vocabulary and set of definitions are applicable to all geoecosystems, they will be rehearsed here.

Scale

Scale refers to the spatial or temporal dimension of a system. The scale of a geoecosystem is determined by an observer according to the problem he or she is interested in (cf. T. F. H. Allen and Starr 1982). A scale must be selected appropriate to the problem in hand. There is no agreed terminology for describing different scales of geoecosystems. The Delcourts (1988) suggested four space–time domains, as they called them, and adopted the prefixes micro, meso, macro, and mega to describe them (Table 1.2). It would be simpler to refer to these scales as small, medium, large, and very large, but the 'm' words seem to have an irresistible appeal.

Statistical methods may be used to elucidate scales of systems. Suitable techniques include power spectrum analysis, multiscale ordination, and Fourier transformation; fractal analysis is particularly useful (e.g. Krummel *et al.* 1987). Scale is of immense consequence in understanding geoecosystems. The structure, function, and dynamics of landscapes depend on scale. Processes and patterns important at one scale may be unimportant at another – the relative importance of controlling variables shifts as scale changes (Meentemeyer and Box 1987). This is illustrated by the case of litter decomposition which, at a microscale, is determined largely by the properties of the litter and the decomposer community, but, at macro- and megascales, is determined mainly by climatic variables (Meentemeyer 1978, 1984, 1989). Likewise, evapotranspiration is determined by vapour-pressure deficit and stomatal processes at the scale of a single leaf or plant, but is driven by net radiation at regional scales (Jarvis and McNaughton 1986). In the same vein, different processes are invoked to explain the distribution of oak seedlings at different scales (Neilson and Wullstein 1983). At a local scale, increased precipitation leads to a decrease in seedling mortality; at regional scales, drier latitudes are associated with the lowest seedling mortality rates. The salient point is simple yet of monumental import: the results of an investigation of a geoecosystem will be influenced by the scale chosen to study the patterns and processes. Fortunately, several studies suggest that, in mesoscale

17

landscapes, relatively few variables are required to predict geoecological patterns, the spread of disturbances, or ecosystem processes such as net primary production or the distribution of soil organic matter. The explanatory power of a set of variables at different scales may be probed, either by using an analytical technique such as regression and varying the grain or extent of the analysis, or by adopting theoretical methods (e.g. Gardner *et al.* 1989, 1992; O'Neill *et al.* 1991).

Resolution

Landscape ecologists use fine scale to refer to minute resolution or small study area, and broad scale to refer to coarse resolution or large study area. This is contrary to the usage in cartography where large scale adverts to high resolution and small scale to low resolution. Besides resolution, landscape ecologists differentiate between grain and extent. Grain is the finest level of resolution possible with a given set of data. It would, for instance, be the pixel size for raster data. Extent is the size of the study area or the duration of the study.

Level of organization

This refers a system's position within an ecospheric hierarchy. In this book, attention focuses on the biological, pedological, and geoecological hierarchies.

There are at least two ways of slicing the biosphere into hierarchical units. First, a genealogical hierarchy can be recognized running from genes and chromosomes, through genomes, demes, incipient species, and species, to monophyletic taxa and all life. Second, a societal hierarchy can be recognized running from individual organisms, through local communities, communities, and biomes, to zonobiomes and the biosphere. A biome is a biotic community considered as a whole, the combined communities of plants and animals (Clements and Shelford 1939). The equivalent term for plants is a formation; and a formation-type is equivalent to a zonobiome. Communities of animals are called communities, though it would be helpful if somebody were to invent the term 'faunation', as an animal equivalent of vegetation. Heinrich Walter (1985) divided the Earth's vegetation into nine zonobiomes (zonal biomes), each corresponding to a genetic climatic type (Figure 1.3 and Plate 1.1).

Many hierarchies of soils have been proposed. A useful one, constructed from a functional consideration of the soil system, runs from soil horizons, through soil tesserae, to soil landscapes and the pedosphere. A different ilk of pedological hierarchy may be erected by regarding pedons (three-dimensional units of almost identical soil profiles, normally between 1 to 10 m² in extent) as individuals that differ from one another in varying degrees and classifying them accordingly. Like pedons form a polypedon, a higher

18

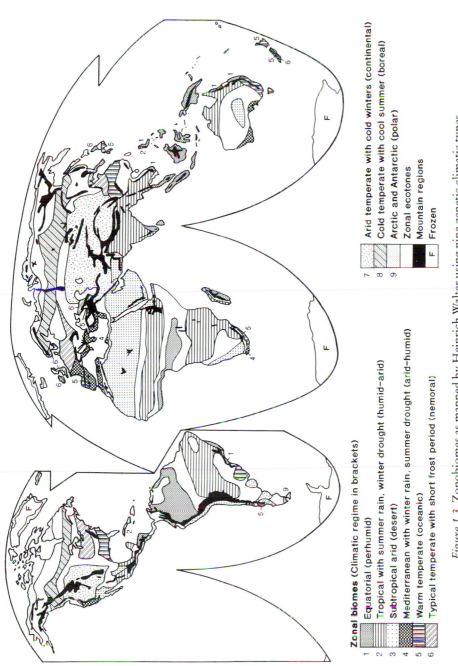

Zonal biomes (Climatic regime in brackets)

1 Equatorial (perhumid)
2 Tropical with summer rain, winter drought (humid–arid)
3 Subtropical arid (desert)
4 Mediterranean with winter rain, summer drought (arid–humid)
5 Warm temperate (oceanic)
6 Typical temperate with short frost period (nemoral)

7 Arid temperate with cold winters (continental)
8 Cold temperate with cool summer (boreal)
9 Arctic and Antarctic (polar)

   Zonal ecotones
   Mountain regions
F Frozen

Figure 1.3 Zonobiomes as mapped by Heinrich Walter using nine genetic climatic types
Source: After Walter and Breckle (1985: fig. 5)

Plate 1.1a Vegetation within Walter's zonobiomes. Humid tropical (equatorial) zonobiome: tropical rain forest in the Danum Valley, Sabah.
Photograph by Ian Douglas

Plate 1.1b Vegetation within Walter's zonobiomes. Seasonal tropical zonobiome: savanna vegetation in the Medway area, Queensland, Australia.
Photograph by Ian Douglas

Plate 1.1c Vegetation within Walter's zonobiomes. Subtropical arid (desert) zonobiome: hot desert vegetation, Great Eastern Erg, Southern Tunisia. Photograph by Keith Sutton

Plate 1.1d Vegetation within Walter's zonobiomes. Mediterranean zonobiome: sclerophyllous woody vegetation (matorral), Rio Aguas Valley, south-east Spain. Photograph by Keith Sutton

Plate 1.1e Vegetation within Walter's zonobiomes. Warm temperate (maritime) zonobiome: temperate evergreen forest (mixed broad-leaf and podocarp species), west coast of New Zealand.
Photograph by Brian Kear

Plate 1.1f Vegetation within Walter's zonobiomes. Typical temperate (nemoral) zonobiome: broad-leaved deciduous forest, mainly oak (*Quercus* spp.) and hornbeam (*Carpinus betulus*), in autumn, Northaw Great Wood, Hertfordshire, England.
Photograph by Richard Huggett

Plate 1.1g Vegetation within Walter's zonobiomes. Arid temperate (continental) zonobiome: tussock grassland in central Otago, New Zealand.
Photograph by Brian Kear

Plate 1.1h Vegetation within Walter's zonobiomes. Cold temperate (boreal) zonobiome: lodgepole pine (*Pinus contorta*) forest in the Canadian Rockies.
Photograph by Brian Kear

Plate 1.1i Vegetation within Walter's zonobiomes. Arctic and Antarctic polar
zonobiome: tundra vegetation at about 900 m, Okstindan, Norway.
Photograph by Wilfred H. Theakstone

level three-dimensional unit that usually corresponds to a soil series. By
grouping like polypedons, and so forth, a classification of soils is erected. The
resulting hierarchical soil taxonomies are akin to the genealogical hierarchy
of organisms, though they do not bear the same evolutionary connotations
– soils are not descended from a common ancestor! Soil classification
schemes are legion, unashamedly nationalistic, and use nomenclature that is
notoriously confusing to the uninitiated. Old systems were based on
geography and genesis. They designated soil orders as zonal, intrazonal, and
azonal; divided these into suborders; and then subdivided the suborders into
Great Soil Groups such as tundra soils, desert soils, and prairie soils. Newer
systems give more emphasis to measurable soil properties that either reflect
the genesis of the soil or else affect its evolution. The most detailed and
comprehensive new classification was prepared by the Soil Conservation
Service of the US Department of Agriculture and published, after many
approximations, in 1975. To ease communication between soil surveyors, the
nomenclature eschewed the early genetic terms and, for units above the series
level, used names derived mainly from Greek and Latin. The taxonomy was
based on class distinction according to precisely defined diagnostic horizons,
soil moisture regimes, and soil temperature regimes. Ten orders were
distinguished: Alfisols, Aridisols, Entisols, Histosols, Inceptisols, Mollisols,

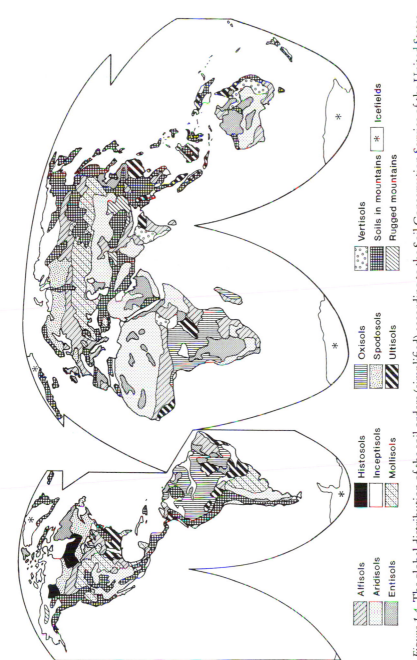

Figure 1.4 The global distribution of the soil orders (simplified) according to the Soil Conservation Service of the United States Department of Agriculture. For explanation and subdivisions see Soil Survey Staff (1975)

Alfisols
Aridisols
Entisols

Histosols
Inceptisols
Mollisols

Oxisols
Spodosols
Ultisols

Vertisols
Soils in mountains
Rugged mountains

Icefields

Oxisols, Spodosols, Ultisols, and Vertisols (Figure 1.4). The orders were successively subdivided into suborders, great groups, subgroups, families, and series. Recently, an eleventh order has been added – Andisols (soils in which more than half the parent mineral matter is volcanic ash).

A hierarchy of geoecosystems may be constructed along the following lines. The smallest unit is the tessera, or landscape element, which is normally a cubic metre or less in volume. Tesserae combine to form mesoscale landscapes with areas up to about 10,000 km². Mesoscale landscapes join to create macroscale landscapes with areas up to about 1,000,000 km², which is about the size of Ireland. In turn, macroscale landscapes are part of megascale landscapes – regions more than 1,000,000 km² in extent that include continents and the entire land surface of the Earth. Landscape ecologists single out three important characteristics of landscapes (Forman and Godron 1986: 9): a particular landscape is influenced by the same macroclimate, by similar features of geomorphology and soils, and by a similar set of disturbance regimes. These characteristics apply to mesoscale landscapes, but not to macroscale landscapes or megascale landscapes. It is not clear if there are natural units in the landscape that correspond to these different scales. There are two polar possibilities. First, landscapes may change gradually as scale is increased and transitions across scales are continuous and imperceptible. Second, landscapes may remain approximately constant as scale is increased and then, at some threshold scale, jump to a new pattern. Current thinking, especially in landscape ecology, is that the second alternative is the more likely, but the issue is far from being resolved.

SUMMARY

The outer Earth may be viewed as a set of coacting spheres – lithosphere, atmosphere, hydrosphere, toposphere, pedosphere, biosphere, and ecosphere. The first three are the spheres of rock, air, and water. The toposphere is the 'relief sphere', the topographic surface lying at the lithosphere–atmosphere and lithosphere–hydrosphere interfaces. The pedosphere, or soil sphere, is defined as that part of the lithosphere influenced by the ecosphere. It may be divided into two parts: the edaphosphere, or what soil scientists might call the proper soil sphere, involving the solum (A and B horizons of a soil profile); and the debrisphere, the part of the pedosphere lying below the edaphosphere. The biosphere is taken to mean the global totality of living things, and the ecosphere is taken to be the global ecosystem, or life plus life-support systems – the biosphere, atmosphere, hydrosphere, pedosphere, and toposphere. The terrestrial portion of the ecosphere is the geoecosphere or landscape sphere. The geoecosphere may be viewed as a hierarchically arranged set of dynamic spatial structures, each of which is a geoecosystem (landscape system).

FURTHER READING

References on the terrestrial spheres are legion. This superabundance of information results from the current popularity of studying things global – global ecology, global biogeochemical cycles, global warming, global climate, global change. Most of the information pertains to store sizes and fluxes in the ecosphere. Readers wishing to follow up this aspect of terrestrial spheres should select appropriate titles from the long list of SCOPE (Scientific Committee on Problems of the Environment) volumes published by John Wiley & Sons. These are a good starting point for information on the major mineral cycles. Books summarizing findings have appeared recently. An example is *Global Biogeochemical Cycles* (Butcher *et al.* 1992). A journal devoted to the subject – *Global Biogeochemical Cycles* – was established in the late 1980s. Discussions of global ecology and the ecosphere would be incomplete without mention of the Gaia hypothesis. For those who do not yet know about Gaia, try James E. Lovelock's books, but be sure to read *Scientists on Gaia* (Schneider and Boston 1991) and *Life as A Geological Force: Dynamics of the Earth* (Westbroek 1991). For a truly stimulating read, try Brian Aldiss's *Helliconia* trilogy – *Helliconia Spring* (1982), *Helliconia Summer* (1983), and *Helliconia Winter* (1985) – a work of science fiction that explores possibilities of the Gaia hypothesis in a magnificent and gripping way. For information on the geoecosystems studied by landscape ecologists an excellent starting point is the textbook by Forman and Godron (1986). A regular perusal of the journal *Landscape Ecology* will help readers keep abreast of developments in the field. A wider perspective is taken by Allen and Hoekstra in their excellent *Toward a Unified Ecology* (1992). The megascale and macroscale patterns of soils are described in *World Soils* (Bridges 1978) and in *Soil Genesis and Classification* (Buol *et al.* 1980). Zonobiomes and zonoecotones are lovingly described in Walter's many books, of which *Vegetation of the Earth and Ecological Systems of the Geo-Biosphere* (1985) is a good place to begin.

2

INTERDEPENDENCE IN GEOECOSYSTEMS

THE 'CLORPT' EQUATION

Jenny's classic 'clorpt' equation provides a conceptual and analytical framework for studying interactions between components of the geoecosphere. It expresses the interdependence of life, soils, climate, rocks, and relief. Ideas about the interdependence of environmental factors were mooted in the late eighteenth century, principally by Johann Reinhold Forster and Alexander von Humboldt. Largely owing to the stimulus provided by Vasilii Vasielevich Dokuchaev, they developed fast during the late nineteenth century. Dokuchaev, a Russian geologist turned pedologist, surveyed large stretches of the chernozems underlying the Russian steppes. This work led him in 1879 to express the view that soil is an independent object and not simply a geological formation; it is a surficial body of mineral and organic substances, produced by the combined activity of animals and plants, parent material, climate, and relief (Joffe 1949: 17). Here was the first categorical statement of the factors of soil formation and, by implication, a basis for exploring in a formal way the connections between ecosystems and their environmental influences. The terrestrial spheres are not specifically mentioned in this formulation, but they are there by implication: climate involves the atmosphere and hydrosphere; animals and plants (plus the three kingdoms of micro-organisms) are the biosphere; parent material is connected to the lithosphere; and relief is part of the toposphere.

During the opening decades of the twentieth century, Frederic E. Clements promulgated influential views about the tight relationship between vegetation and environmental factors. Clement's views were holistic and held up climate as the key to understanding vegetation distribution. None the less, Clements allowed that soil and terrain could also exert some influence on vegetational development. In brief, he envisaged a five-stage sequence, or sere, starting with pioneer plants colonizing a new area and ending with a climax community becoming established. He called the seres leading to climatic climax vegetation types priseres. The course of a prisere reflects the nature of the initial conditions. Two contrasting types were recognized –

xeroseres and hydroseres. As the name suggests, xeroseres begin development under dry conditions, either on bare rock surfaces (lithoseres) or on moving sands (psammoseres). Hydroseres begin in parts of the landscape covered by water. Fresh water hydroseres are associated with lakes and swamps, while haloseres are found on estuarine flats. During all priseres, the community of plants becomes less and less controlled by soil and terrain and more by climatic factors. Clements incorporated the effect of relief as an 'arresting' factor: the extreme steepness or flatness of the land may prevent a prisere from attaining a climatic climax state and is held back, or arrested, in a subclimax condition. This, it has been suggested, is the case in the Fens of England where, though climate would allow the establishment of oak woodland, hydrological conditions associated with very flat, low-lying land prevent the germination of trees and maintain the vegetation as carr dominated by alder (*Alnus glutinosa*) and willows (*Salix* spp.).

In pedology, the first major elaboration of the state-factor approach initiated by Dokuchaev was due to Chas F. Shaw (1930). Shaw argued that soils are formed by the modification, and partial decomposition and disintegration, of parent material owing to the action of water, air, temperature change, and organic life. He expressed soil formation according to the formula:

$$S = M (C + V)^T + D$$

which states that soil, S, is formed from parent material, M, by the work of climatic factors, C, and vegetation, V, over a time, T, but the process may be modified by erosion of, or deposition upon, the soil surface, D. Shaw noted that each of the factors in soil formation is important in determining the character of soil, though under local conditions any one factor may exert a dominant influence. At around the same time, the German botanist Reinhold Tüxen (1931/32) recognized a complex of interacting factors that influence one another: rocks, water, climate, relief, soils, plants, animals, and Man (Figure 2.1). This view of Nature was later to find expression in the idea of the holocoenotic environment (Allee and Park 1939; Billings 1952).

The relations between soils and the other terrestrial spheres were worked out by Hans Jenny. His early ideas seem first to have been set down in 1930. He approached soils from a general theory of state, in which 'soil properties, soil processes, and soil-forming factors are united into a comprehensive system' (Jenny 1930: 1053). His most famous and lasting contribution was the 'clorpt' equation (Jenny 1941). This suggests that any soil property, s, is a function of soil-forming factors:

$$s = f(cl, o, r, p, t, \ldots)$$

where cl is environmental climate; o is organisms (the fauna and flora originally in the system and that entering later); r is topography, also including hydrological features such as the water table; p is parent material,

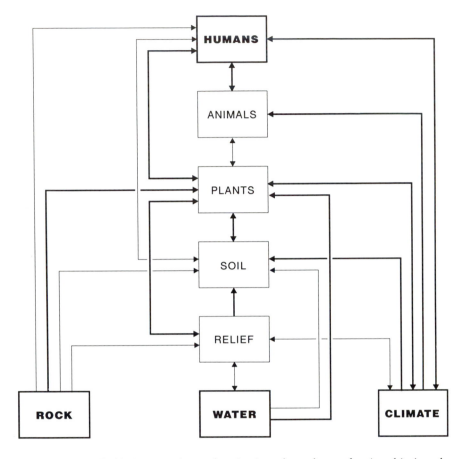

Figure 2.1 Reinhold Tüxen's scheme showing interdependence of various biotic and abiotic environmental factors
Source: After Tüxen (1931/32)

defined as the initial state of soil when pedogenesis starts; t is the age of the soil, or absolute period of soil formation; and the dots are additional factors such as fire. The 'clorpt' equation served as an effective tool of research for many decades. It was extended by Jack Major (1951), and later by Franklyn Perring (1958), to embrace the entire ecosystem – soils, vegetation, and animal life. Jenny (1961, 1980) offered his own extension that included ecosystems (entire sections of landscapes). He derived a general state-factor equation of the form:

$$l, s, v, a = f(L_0, P_x, t)$$

where l is ecosystem properties such as total carbon content, primary production, and respiration; v is vegetation properties such as biomass, species frequency, and sodium content; a is animal properties such as size, growth rate, and colour; and s is soils properties such as pH, texture, humus content; L_0 is the initial state of the system, that is, the assemblage of properties at time zero when development starts (the L stands for the ecosystem, or *l*arger system, of which the soil is part); P_x are external flux potentials; and t is the age of the system. The state factors are groups of variables associated with L_0 and P_x. The initial state of the system is defined by parent material, p, and by the original topography and water table, r. The external flux potentials are environmental properties that lead to additions and subtractions of matter and energy to and from the system. They include environmental climate, cl, a biotic factor, o, comprising fauna and flora as a pool of species or genes, active or dormant, that happen to be in the ecosystem at time zero or that enter it later. The biotic factor is thus distinct from the vegetation that grows as the system develops; this appears as a system property on the left-hand side of the equation. Other external fluxes would include dust storms, floods, and the additions of fertilizers; these could be given symbols and entered separately in the equation if so desired. In an extended form, the general state-factor equation becomes:

$$l, s, v, a = f(cl, o, r, p, t, \ldots)$$

Which brings us back to the 'clorpt' equation, only this time it applies to ecosystems and not just soils. The latest version of the 'clorpt' equation considers the place of the human species in the state-factor theory of ecosystems (Amundson and Jenny 1991).

Jenny (1980: 203) said that the 'clorpt' equation is a synthesis of information about land ecosystems, or what in this book are referred to as geoecosystems. He suggested that, in favourable landscapes, the factors could be sorted out and assessed as six groups of idealized ordinations which revealed the effect of one state factor on a single ecosystem property, all other factors being held constant. Today, ordination would be used to extract gradients from the data, either directly or indirectly (R. H. Whittaker 1967; Shimwell 1971: 235–277), though they would probably not be composed of single factors – a gradient would be composed of a mixture of closely related state factors. Gradients for different factors would be dominant or subordinate in different cases. The resulting relationship between an ecosystem property and a dominant environmental factor could be expressed as a mathematical function of some kind, usually a linear or curvilinear regression equation. Given five state factors, Jenny proposed five broad groups of functions or sequences: climofunctions or climosequences, biofunctions or biosequences, topofunctions or toposequences, lithofunctions or lithosequences, and chronofunctions or chronosequences. He also included dotfunctions and dotsequences to allow for the effects of other factors such as fire.

Jenny's formulation is exceedingly valuable, especially as a conceptual tool. It is still much cited and used, albeit in slightly modified forms, by pedologists and ecologists (e.g. Van Cleve and Yarie 1986; Van Cleve *et al.* 1991). A drawback with the 'clorpt equation' is that it cannot be readily rendered into a form compatible with dynamical systems theory. Nor for that matter can any other formula of soil development offered over the last fifty years. The formulae suggested by Sergius Alexander Wilde (1946) and Donald Lee Johnson and Thomas K. Rockwell (1982), and to a lesser extent that offered by Johnson and Donna Watson-Stegner (1987; D. L. Johnson *et al.* 1990), could, with some basic adjustments, be expressed as a set of dynamic system equations. Wilde (1946: 13) recast Dokuchaev's formula so that time was expressed as a differential:

$$s = \int (g, e, b) \, dt$$

where s is soil, g is geological (parent) material, e is environmental influences, and b is biological activity. This model was further modified by Charles George Stephens (1947) who split environmental influences into climate, c, relief, r, and water table, w, and changed the g and b factors into parent material, p, and organisms, o:

$$s = \int f(c, o, r, w, p) \, dt$$

which is almost identical to the 'clorpt' equation save in that time is seen to influence all factors.

THE 'BRASH' EQUATION

A reformulation inspired by Jenny's 'clorpt' equation was made independently by the present author (Huggett 1991) and Jonathan D. Phillips (1993c). However, rather than rejigging the 'clorpt' formula, it may be more rewarding to formulate a model describing interdependence among the terrestrial spheres by starting with the spheres themselves.

A model of geoecosystems

At the most general level, the entire ecosphere may be thought of as a system consisting of a huge number of interrelated elements, each of which can be measured in many ways. Measures would include things such as mass, length, acidity, species diversity, and birth rates. Each measure describes a state of some component of the ecosphere, and is known as a state variable. In the most general case, the change in a state variable is a function of change in itself, a change in all other state variables within the system, and the effect of any external variables acting on the system. It follows that, if the system be fully interconnected, a change in one state variable has the potential to effect a change in all the state variables comprising the system. Let us express these ideas mathematically. The state of the ecosphere at time $t + 1$, e_{t+1}, depends on two things. First, it depends on its previous state, e_t. Second, it depends on driving or forcing variables, z, acting on the system from outside its boundaries. These assumptions may be expressed as

$$e_{t+1} = e_t + z$$

Writing this equation as a rate of change gives

$$\frac{de}{dt} = e + z$$

This equation is a very general statement of ecospheric dynamics. Equations like it, admittedly more fully developed, are used to predict the overall dynamics of the ecosphere. A case in point is the Daisyworld model designed to demonstrate Gaian temperature regulation (Watson and Lovelock 1983).

The basic equation may be elaborated by taking separate account of each terrestrial sphere. Now, there is considerable feedback between the terrestrial spheres. It is well known that vegetation cover influences relief development, that relief affects climate at scales ranging from local to global, and so forth. For this reason, it seems logical to define all terrestrial spheres, save the lithosphere, as internal state variables, leaving forces from the cosmosphere and the solid Earth as the driving variables. This idea may be formalized by defining relations between the component spheres. Let the spheres be represented by the following symbols: the biosphere (life) b, the toposphere, r, the atmosphere, a, the pedosphere (soils), s, and the hydrosphere, h. These spheres interact so change in any sphere may be equated as a function of the state of all the spheres plus the effects of the driving variables, z. Writing this symbolically yields

$$\frac{db}{dt} = f(b, r, a, s, h) + z$$

$$\frac{dr}{dt} = f(b, r, a, s, h) + z$$

$$\frac{da}{dt} = f(b, r, a, s, h) + z$$

$$\frac{ds}{dt} = f(b, r, a, s, h) + z$$

$$\frac{dh}{dt} = f(b, r, a, s, h) + z$$

This set of equations describes, in a general way, the dynamics of the ecosphere as a whole. It states that all components in life and life-support systems interact to some extent; they are all interrelated, and they are all influenced by various driving variables external to the ecosphere. For convenience, let us call it the 'brash' equation. The cynical reader may carp that the 'brash' equation is a fancy way of saying that everything is connected to everything else. To an extent that is true, but not all the components would necessarily be included in a particular application, and not all the components that were included would necessarily interact. Indeed, if recent work on complexity in dynamical systems is a secure pointer, the most interesting situations may occur when direct links are relatively restricted, an idea developed a little in the epilogue (for a readable and enjoyable review, try Lewin 1993).

This book is concerned with the landscape sphere, or geoecosphere, which consists of a hierarchy of geoecosystems. Since geoecosystems are the result of coacting processes in the terrestrial biosphere, toposphere, atmosphere, pedosphere, and hydrosphere, they may be described using the 'brash' formula. The advantages of doing so are threefold. First, the 'brash' equation is a mathematically sound representation of the structure, function, and dynamics of geoecosystems; at the same time, it is amenable to analysis by the less rigorous, but often very revealing, battery of statistical techniques designed to probe multivariate situations. Second, it seems to capture the 'ecological' interdependence of geoecosystem components more satisfactorily than the 'clorpt' equation, allowing as it does for reciprocity between all factors. Third, it suggests an evolutionary, as opposed to developmental, view of geoecosystems. Before exploring environmental influences on life and soils in the context of geoecosystems, these points and their implications will be examined.

Dynamical system equations

The 'brash' equation is written as a set of simultaneous differential equations that describes system dynamics at any level of the geoecological hierarchy. As presented above, the 'brash' equation is in its most general form, though it could be more succinctly written using matrix notation as

$$\frac{dx}{dt} = f(x) + z$$

where x is a vector of state variables describing the geoecosystem under study, $f(x)$ is a matrix defining the interactions between state variables, and z is a vector of driving variables. Interaction between system components is defined by parameters, $a_{i,j}$. The full set of parameters may be written as an $n \times n$ interaction matrix, A, which may be added to the equation:

$$\frac{dx}{dt} = f(x, A) + z$$

Systems of equations of this kind, first developed by Ludwig von Bertalanffy (1950), are found in many fields, with the measures used to define the state variables varying from one application to another. They are used in geomorphology (Huggett 1988; Phillips 1993a, 1993b, 1993c), biogeochemistry (Lasaga 1983; Kump and Garrels 1986), population ecology (May 1973; Pimm 1982, 1992; Puccia and Levins 1991), landscape ecology (Kadlec and Hammer 1988), watershed hydrology (Cosby *et al.* 1985), and many other disciplines.

The general stability conditions of dynamical system equations can often be found by appropriate matrix methods (Puccia and Levins 1985, 1991). These solutions are qualitative, only requiring information on positive and negative links within a complex system. Quantitative analysis requires that a set of state and driving variables be selected, that the assumed relationships between them be decided upon, and that variables and parameters be given values. There is plenty of theoretical and empirical work on which to draw in doing this, though calibrating a dynamical systems model is not easy. It would be inappropriate to delve into dynamical systems modelling in this book, as the main concern is with the 'brash' equation as a conceptual tool, as a way of thinking about geoecosystem dynamics. Readers unfamiliar with dynamical systems modelling will find very simple accounts in two of my own books (Huggett 1980, 1993). Advanced coverage can be found in the references cited in the previous paragraph.

Interdependence in the environmental complex

Jenny's general state-factor equation applied to the 'larger system' (an ecosystem) includes animals and plants as system properties, denoted as a and v by Jenny, as well as soil. It provides a useful way of thinking about relations between living things and their environment. The functional–factorial approach adopted by Jenny and others relates soil and biological factors to a set of environmental variables that are assumed to act independently of one another. Or, as Jenny (1980: 202) circumspectly and diplomatically puts it, under some circumstances, the variables describing state factors can be independent of one another. This seems doubtful: climate influences the terrestrial water cycle and, through slope processes, the topography; topography can influence climate; parent material influences topography and the water table, and so forth. And soils and vegetation may influence climate and other ecospheric systems. It makes sense to view the soil as part of a larger system, but the components of the larger system are interrelated and there are problems in regarding them as independent state factors, except under stringently controlled conditions (cf. Crocker 1952). That is not to say that the influence of one component upon another is undetectable. To be sure, soil nitrogen content does vary with mean annual rainfall, soil texture is influenced by parent material, soluble materials are leached from hills and accumulate in valleys. In establishing a climosequence, lithosequence, toposequence, or whatever, the normal practice is to eliminate the effect of other influences by holding them constant. A big problem with this procedure is that a relationship between two variables is teased out of a richly multivariate situation. It appears commonly to be the case that geoecosystem properties are influenced by various factors acting simultaneously, as recognized in the 'brash' equation, a fact that will be disguised in univariate relations. This is something to remember when looking at environmental influences on biospheric and pedospheric evolution. To be fair, Jenny (1980) did acknowledge this problem and met it head on by experimenting with a multivariate statistical model (e.g. Jenny et al. 1968). However, rather than sticking doggedly to the notion of state-factor independence, and ironing out any uncomfortable dependencies with sophisticated statistical techniques, it is perhaps better to take interdependency on board and revamp the 'clorpt' equation and all its derivative formulae.

To appreciate the difference between the 'clorpt' and 'brash' formulations, the soil component of the 'brash' equation may be scrutinized:

$$\frac{ds}{dt} = f(b, r, a, s, h) + z$$

This is, in essence, a version of the 'clorpt' equation with time included as a derivative. Biosequences, climosequences, and so on, could still be established, providing the effects of individual terms on the right-hand side were investigated with all other terms kept constant, which practice is adhered to by users of the 'clorpt' formula. The problem with doing this is that the soil-system equation is one of a set, all component equations of which operate simultaneously and describe the dynamics of a full geoecosystem. Singling out one state variable might produce erroneous results because the system acts as a whole, often in ways that cannot be predicted from the dynamics of individual parts. This is why great care should be taken when interpreting relationships between single soil and ecosystem properties and individual state factors. None the less, providing all investigators have a badge pinned to them cautioning that 'the whole is greater than the sum of the parts', then climofunctions, lithofunctions, and the rest could be explored within the context of the 'brash' formula.

The role of driving variables, z, in the 'brash' equation needs some clarification. Driving variables are, by definition, external to a system under investigation. However, internality and externality are scale dependent. To appreciate this point, it may help to consider a geoecosystem, as described in the basic 'brash' equation, as an interdependent set of dynamic fields. Some of the fields are continuous, some discontinuous, some patchy. Climate consists of many atmospheric and hydrospheric fields, important among which are temperature fields, precipitation fields, evaporation fields, and wind fields. Topography may be deemed a potential energy field that enables material in the debrisphere and hydrosphere to move downslope under the force of gravity. Rock is a lithological field, often patchy in nature, that varies in composition and structure. The biosphere is a complex of fields that is patchy at small scales but continuous on a global scale. It has to adapt to, and harness, the dynamic environmental fields. The dynamic fields that constitute a geoecosystem are themselves set in wider ecospheric fields. In turn, the ecospheric fields are set within dynamic geological and cosmic fields. Thus, for example, topography is an interactive part of the ecosphere that is itself disturbed by geological, and to a lesser extent cosmic, forces. It may be thought of as a spatial field that influences, and that is influenced by, fluxes in the biosphere, debrisphere, and hydrosphere. The suggestion that environmental fields are a useful way of looking at influences on geoecosystems extends the idea of environmental gradients, so convincingly shown by Paul A. Keddy (1991) to be a powerful and overlooked research tool, to three dimensions. Gradients are parts of environmental fields.

Interaction between the dynamic fields of a geoecosystem can be viewed as the disturbance of one system by another. From the point of view of the biotic portion of a geoecosystem, the atmosphere, toposphere, hydrosphere, and pedosphere are external variables and potential disturbers of the fauna and flora. For example, plants in a geoecosystem adjust to the changing topographic field. At the mesoscale, this results in vegetation catenae. Interactions of systems within the biotic landscape may involve the disturbance of one biotic field by another, as when grassland is grazed. Disturbance occurs at all spatial scales, from the dung beetle burrowing into the soil, through buffalo wallows and the loss of tree stands, to the destruction of entire biota.

It is plain from the above discussion that driving variables are processes or forces originating outside a system that inflict disturbance. These disturbances may be harnessed to the system's advantage, as with solar radiation, or the system may have to accommodate them, if possible. This is why it is so difficult to distinguish between disturbing variables and driving variables. Consider the case of disturbance by fire. Timothy F. H. Allen and Thomas W. Hoekstra (1992: 82) explain that individual fires destroy and disturb susceptible communities. If the fire regime persists, the vegetation will become adapted to fire. At that point, the long-term survival of the community requires that fire dispenses with invaders that are not fire-adapted. So, the fire-adapted community has integrated fire as part of the system. Paradoxically, fire disturbance destroys biomass of individuals but sustains the community of which the individuals are part. The same argument could apply to other disturbing factors in geoecosystems.

An evolutionary view of geoecospheric dynamics

Inherent in the 'brash' formulation is an evolutionary view of geoecosystems, and of the biosphere and pedosphere. To explain what is meant by this and to show its significance, it is perhaps best first to describe the opposing view encapsulated in the factorial–functional model.

Modern ecology has inherited from the late nineteenth and early twentieth centuries a developmental view of soils and vegetation that may, ultimately, be the gift of Charles Darwin, but has been handed down from William Morris Davis, Dokuchaev, and Clements. The argument runs that soil forms or develops progressively under the influence of the state factors. Eventually, the soil will be in equilibrium with the prevailing environmental conditions and change no more – it will be a mature soil. Likewise, vegetation develops through successive changes until it attains a mature or climax state that is in balance with the environmental, and especially climatic, conditions. As hypotheses, these ideas are sound enough. The big problem is that a mountain of evidence points to environmental conditions being inconstant. Considering this fact, it is improbable that a developmental sequence of soils or vegetation will run its course under a constant environment. And, to complicate the picture even more, it has been found that most ecospheric systems are best viewed as dissipative structures replete with non-linear relations and forced away from equilibrium states by driving variables. This non-linearity in systems removed from equilibrium may generate chaotic regimes in which internal system relations and thresholds drive systems through a series of essentially unpredictable states that are strongly dependent on the initial conditions. This is in complete contrast to the developmental view of soils and vegetation wherein the initial state is thought to be of little importance, save in exceptional circumstances. Environmental inconstancy and non-linear dynamics lead then to a far more dynamic picture of biospheric and pedospheric change than an early generation of ecologists could scarcely have imagined in their most fanciful dreams. The systems of the biosphere and pedosphere are generally plastic in nature and respond to changes in their environment and to thresholds within themselves. The result is that soils and vegetation evolve, rather than develop: their genesis involves continual creation and destruction at small, medium, large, and very large scales, and may progress or retrogress depending on the environmental circumstances; there is not of necessity a predetermined developmental path that they pursue willy-nilly. Interestingly, Evelyn C. Pielou (1977) commented many years ago that environmental gradients are more worthy of study than the chimerical homogeneous environments so much admired by ecologists. Ever-changing and mutually dependent fields and gradients are part of the evolutionary conception of geoecosystems.

This evolutionary view of ecospheric systems is of the utmost significance and applies to geoecosystems at all levels and scales. It means that at any instant, all geoecosystems are unique and changing, and are greatly influenced by historical events (cf. Bennett 1993). Defining geoecosystems in this way means that predicting change is difficult, even though the relationships expressed in the 'brash' equation are deterministic. Vegetation and soils formed under the same environmental constraints are likely to be broadly similar but will invariably differ in detail.

An evolutionary model of the soil was first developed by Johnson and Watson-Stegner (1987). It was an attempt to allow for the fact that soil evolves in an ever-changing environment so that polygenetic soils are the norm. Its keynote is polygenesis and stands in antithesis to monogenetic models and notion of zonal soils, normal soils, and climax soils. It is summarized by the equation

$$s = f(P, R)$$

where s is soil, P is progressive pedogenesis and includes process, factors, and conditions that promote differentiated profiles, and R is retrogressive pedogenesis and includes processes, factors, and conditions that promote simplified profiles. There are problems with this model that are not met with in the 'brash' formulation. While not denying that soil evolution could be regressive or progressive, it is unduly complicated to single out processes that are deemed retrogressive from those are deemed progressive. One reason for this is that the designation of a process as progressive or retrogressive may well vary with the scale of the system (pedon, catena, soil landscape). By contrast, the 'brash' model automatically caters for positive and negative effects in the system. A related evolutionary model was developed (or possibly evolved) by Johnson and his colleagues (1990). In summary, it assumes that

$$s = f(D, P, \frac{dD}{dt}, \frac{dP}{dt})$$

where s is a soil property or the degree of pedogenesis, D is a set of dynamic vectors and dD/dt their rate of change through time, P is a set of passive vectors and dP/dt their rate of change through time. The dynamic vectors include energy fluxes, mass fluxes, the frequency of wetting and drying events, organisms, and pedoturbation. The passive vectors include parent material, the chemical environment of the soil, permanently low water tables, the stability of slopes, and pedogenetic accessions such as fragipans, natric horizons, and histic horizons. This model suffers a similar defect to its predecessor in so far as the designation of dynamic and passive factors can be made only for a certain scale of investigation. In addition, many of the so-called passive factors change with time and are interrelated. This overall dynamism of soils and the environmental complex is allowed for in the 'brash' equation.

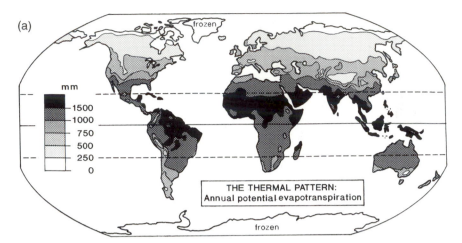

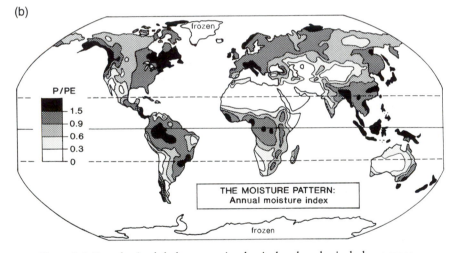

Figure 2.2 Four basic global patterns in physical and ecological phenomena.
(a) Thermal pattern, as represented by annual potential evapotranspiration.
(b) Moisture pattern, as represented by the ratio of annual rainfall to potential
evapotranspiration (*P/PE*). (c) Throughput pattern, as represented by net primary
productivity. (d) Accumulation pattern, as represented by carbon stored in
undisturbed soils
Source: After Box and Meentemeyer (1991)

THE GLOBAL SETTING

It is reasonable to suggest that environmental influences on life and soils
within geoecosystems depend on scale. Much evidence suggests that climate
is the chief determinant of geoecological processes in megascale landscapes
(where the climatic fields consist of latitudinal and longitudinal gradients)

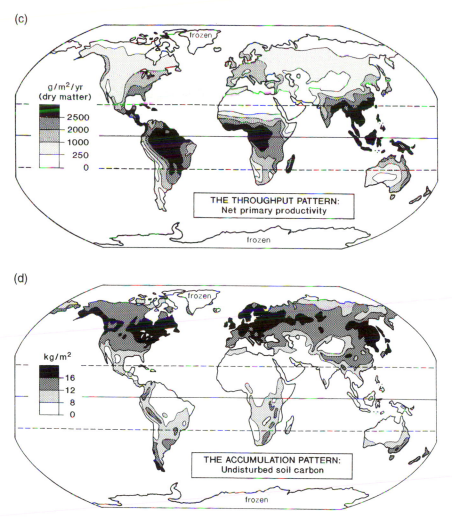

Figure 2.2 continued

and in mountainous macroscale landscapes (where the climatic fields consist of altitudinal gradients). Topography (aspect, slope, landscape position, and insularity) and parent material appear to have a predominating influence in mesoscale landscapes. That is not to deny that all environmental factors may make themselves felt at all scales: topography can influence global climates and thus, indirectly, influence large-scale landscapes; microclimates affect small-scale landscapes. The point is that climate appears to exert a very strong influence at the global level. It does so largely by dictating the supply of energy and water at the Earth's surface. Heat and water are the chief drivers of virtually all forms of energy and matter transfer occurring in geoecosystems. Production and consumption in ecosystems, adaptation in animals and plants, and soil genesis are all strongly influenced by energy and the availability of water. Other climatic variables, such as wind speed, snow cover, lightning strike frequency, and sunshine hours, influence geoecosystems, but temperature and available moisture can be considered the master driving variables.

Given the tight constraints placed by climate on many geoecospheric processes, it is not surprising that many important phenomena of geoecosystems, when mapped on a global scale, display four basic patterns: a thermal pattern, a moisture pattern, a throughput pattern, and an accumulation pattern (Figure 2.2a–d). These patterns are largely dictated by zonal and regional climatic gradients (Box and Meentemeyer 1991). As they seem so important in explaining many features of geoecosystems and provide a sort of backdrop against which the influence of other environmental factors should be viewed, it seems sensible to describe them.

Thermal pattern

The thermal pattern is related to the amount of energy entering geoecosystems. It mirrors energy availability and is largely determined by the Earth's surface radiation or energy balance. Geoecological phenomena that follow the thermal pattern are driven by solar energy and are not generally limited by moisture availability. Consequently, they show their highest levels at the equator and lowest levels at the poles (Table 2.1). Examples are insolation, temperature levels, annual potential evapotranspiration, and community respiration rate.

Moisture pattern

The moisture pattern is related to the amount of precipitation entering geoecosystems. It mirrors water availability. Available moisture is far more important in influencing geoecological processes than precipitation totals alone, since it is a measure of the water that passes through the landscape (roughly the precipitation less the evapotranspiration). This point is readily understood with an example: a mean annual rainfall of 400 mm might support a forest in Canada, but a dry savanna in Tanzania. Geoecological phenomena adhering to the moisture pattern include precipitation, the ratio of precipitation to potential evaporation, vegetative cover, and solar energy efficiency of net primary production. They tend to have high or medium values in the tropics and temperate belt and low values in the arid subtropics and continental temperate belts (Table 2.1).

The sheer potential of the water cycle to influence geoecosystems is made evident by considering the turnover of atmospheric moisture. Although the

Table 2.1 The expression of the four ecological patterns in the main climatic zones

Climatic Zone	Thermal pattern	Moisture pattern	Throughput pattern	Accumulation pattern
Equatorial with a humid, diurnal climate (mean daily temperature fluctuation greater than mean temperatures of warmest and coldest months)	High	High	High	Low to high
Tropical with summer rains, winter drought	High	Medium	High	Low to medium
Subtropical arid	Medium	Low	Low	Low
Intermediate with winter rain and summer drought (Mediterranean)	Medium	Medium	Medium	Medium
Warm temperate with almost no cold winter season and very wet summers (oceanic west coast)	Medium	High	Medium	High
Warm temperature (east coast)	Medium	High	Medium to high	Medium
Typical temperate with short period of frost (nemoral)	Medium	High to medium	Medium	Medium to high
Arid temperate with a cold winter and dry summer (continental)	Medium	Low	Medium to low	Low to high
Temperate arid with a marked drought period and short wet season (continental)	Medium	Low	Low	Low
Cold temperate with cool summer (boreal)	Low	High to low	Low to medium	High
Arctic and Antarctic (polar)	Low	High to low	Low	High to low

Source: After Box and Meentemeyer (1991)

atmospheric store of water is but a minute part of the hydrosphere, its significance to the other spheres is disproportionately large. The volume of water stored in the atmosphere is 13,000 km³. Since the area of the Earth's surface is 510,000,000 km², it is a matter of simple arithmetic to work out that, if all the water vapour in the atmosphere were to condense, it would form a layer 2.54 cm, or exactly one inch, deep. Globally, the mean annual precipitation is 97.3 cm. So, there must be 97.3/2.54 ≈ 38 precipitation cycles per year, and the average life of a water molecule in the atmosphere is therefore 365/38 ≈ 10 days. Furthermore, the global store of surface fresh water, if not replenished, would be lost by evaporation in as little as five years, and drained by rivers in ten.

Throughput pattern

The throughput pattern mirrors the simultaneous availability of heat and moisture. Geoecological phenomena conforming with this pattern include actual evapotranspiration, soil texture, the field capacity of soils, net primary production, gross primary production, litter production, the decomposition rates of wood and litter, soil acidity, the base saturation of soils, and possibly plant and animal species richness. They tend to have high or medium values in the tropics, warm temperate zone, and typical temperate zone, and low values in the arid subtropics and continental temperature regions (Table 2.1).

The interaction of energy and water supplies is of enormous importance in understanding geoecological processes. For a plant to use energy for growth, water must be available. Without water, the energy will merely heat and stress the plant. Similarly, for a plant to use water for growth, energy must be obtainable. Without an energy source, the water will run into the soil or run off unused. The significance of the throughput pattern is revealed by considering the water balance of a geoecosystem:

$$\text{Precipitation } (P) = \text{Evapotranspiration } (E) + \text{Runoff } (D) + \text{Storage } (S)$$

In the long term, storage can be taken as constant and ignored in drawing up a global inventory of annual water fluxes. The water balance of a geoecosystem determines three things. First, it establishes how much energy and water are available concurrently to drive geoecological processes. Second, it decides the water deficit, the difference between potential and actual evapotranspiration; that is, how much evaporative demand is not met by available water. Third, it sets the water surplus, that is, how much water leaves the soil by surface or subsurface flow without being evaporated or transpired.

Accumulation pattern

The accumulation pattern mirrors gains of material over long time-spans. Geoecological phenomena complying with this pattern include standing biomass, litter accumulation, and soil carbon content. For instance, soil carbon tends to increase to a steady-state value when gains from leaf-fall balance losses from decomposition. And standing phytomass tends to accumulate so long as annual net primary production exceeds annual litterfall. The levels at which geoecological phenomena conforming to the accumulation pattern are expressed in different climate zones are somewhat variable (Table 2.1).

SUMMARY

Interdependence among the terrestrial spheres in geoecosystems was recognized by Dokuchaev who initiated the state-factor approach to soil evolution. This approach was embellished through the twentieth century. Its highest development is Jenny's latest version of his 'clorpt' equation in which ecosystem properties are expressed as a function of climate, organisms, relief, parent material, and time. Valuable though the state–factor or factorial–functional approach has been, it is not fully compatible with dynamic systems theory. The 'brash' equation proposed in this book reformulates Jenny's classic 'clorpt' equation in the mathematical language of systems theory. The outcome is a general statement of the dynamics of the geoecosphere in which change results from the interplay of factors within the geoecosphere, and from the influences of cosmic and geological factors originating outside the geoecosphere. The 'brash' equation is at once ecological and evolutionary. It is 'ecological' because it allows for reciprocity between all geoecospheric factors – everything is connected to everything else. It is 'evolutionary', as opposed to 'developmental', because it allows that conditions within and outside geoecosystems are ever changing. Internal change partly results from non-linear behaviour of geoecosystem components. External factors, such as solar energy, gravity, and tectonics, are characteristically dynamic. In an evolutionary view, geoecosystems and their components continually respond to changing internal and external circumstances. This is contrary to the classic development view where geoecosystems and their components – especially vegetation and soils – are seen as progressing (developing) towards a mature or climax state within an essentially stable climatic environment.

The rest of the book looks at internal and external environmental influences on life and soils within geoecosystems. The importance of scale in understanding relationships is stressed. Broadly speaking, relationships in megascale and mountainous macroscale geoecosystems are dominated by climatic 'fields', while relationships in mesoscale geoecosystems are primarily determined by topographic and lithological 'fields'. Globally, four ecological patterns, each corresponding to a different type of climatic field, are distinguishable: a thermal pattern, conforming to energy availability at the Earth's surface; a moisture pattern, conforming to precipitation input; a throughput pattern, conforming to water actually moving through a landscape (roughly precipitation less evapotranspiration); and an accumulation pattern, conforming to combinations of climatic fields favouring the long-term accumulation of biomass, litter, and soil materials.

FURTHER READING

Connections between ecospheric components are discussed in the SCOPE volumes, as mentioned in Chapter 1. Dynamic systems models are covered in several texts. A simple account is offered in *Modelling the Human Impact on Nature: Systems Analysis of Environmental Problems* (Huggett 1993). The 'clorpt' equation is clearly described and exemplified in Jenny's *The Soil Resource: Origin and Behavior* (1980). Jenny's first book, *Factors of Soil Formation: A System of Quantitative Pedology* (1941), although published nearly sixty years ago, still merits perusal. Other useful discussions of functional–factorial aspects of soil genesis can be found in John Gerrard's *Soil Geomorphology: An Integration of Pedology and Geomorphology* (1992) and in John A. Matthews's *The Ecology of Recently-Deglaciated Terrain: A Geoecological Approach to Glacier Forelands and Primary Succession* (1992). The 'brash' equation makes its debut in this book. Discussion of the evolutionary view of soils appears in Donald Lee Johnson's recent work.

Part II

INTERNAL INFLUENCES

3

CLIMATE AND SOILS

Soil, an integral part of geoecosystems, is influenced by climate at local, regional, and global scales. Climatic factors prescribe the kind and rate of some key pedogenetic processes (especially weathering processes and soil drainage), strongly influence the vegetation growing in the soil, and, to a lesser extent, influence relief and topography by affecting hillslope and river processes. Connections between climate and soils are usually hard to decipher. There are two reasons for this. First, some soil properties may be more sensitive to extremes of climate than they are to mean conditions. Second, climate is seldom constant for long: climatic change appears to be the norm. For this reason, a shadow of doubt is cast over the validity of the work relating soil type to climate carried out within the classic developmental framework. However, it may pay to see what this work has had to offer before dismissing it out of hand.

ZONAL SOILS

The role of climate as a soil-forming factor was recognized independently in the United States and Russia in the last decade of the nineteenth century. In America, Eugene Woldemar Hilgard carried out extensive studies of soils. He collected a large body of data on the acid soluble constituents of soils in the arid and humid subtropical lands of the southern United States (Hilgard 1892). His studies led him to the conclusions that the soils of a region are intimately related to the prevailing climatic conditions, and that climate tends materially to influence the character of soils, even those formed from the same rocks. In Russia, Dokuchaev, the father of soil science, established the principle of geographical soil types: each soil formation is associated with a definite climatic belt, unlike the underlying rock formations that are unrelated to climate. Nikolai Mikhailovich Sibirtsev took up Dokuchaev's thesis. He unified Dokuchaev's factors of soil formation and established their differential role in soil development (see Joffe 1949). For Sibirtsev, moisture was the primary climatic factor in soil formation, each climatic zone engendering a diagnostic soil type – a zonal soil. In Russia, the importance

49

attached to climate as a soil former reached its extreme expression in Konstantin Dimitrievich Glinka's (1914) opinion that a mature soil, developed under a constant climate and vegetation, will be the same regardless of parent material. The views of the Russian pedologists spread from Europe to the United States where Curtis Fletcher Marbut rhapsodized about soil-forming factors (e.g. Marbut 1927).

The early work on soil–climate relations led to a number of climatic classifications of soils (see Clayden 1982). Successful though they were, these classifications suffered from using ratios of annual precipitation to temperature, evaporation, and humidity, or from using simple annual precipitation and temperature indices. Climatic ratios and indices are readily mapped. Once mapped, it is easy to compare their distribution with the mapped distribution of zonal soil types (see Huggett 1991: 49–54, 112–113). They are not readily related directly to soil processes. Direct relationships between climate and processes that operate in the pedosphere were not investigated until C. Warren Thornthwaite (1931, 1948) developed his water-balance approach to climatic classification. This work was pursued in depth by Rodney J. Arkley (1967) who investigated soils in the western United States. Arkley undertook water-balance analyses of more than a thousand climatic stations. From climatic records, he extracted data on actual evapotranspiration and mean annual temperature, and then computed what he termed the 'leaching effectiveness of the climate' (measured in cm of water). He then plotted the Great Soil Groups associated with each climatic station against the three climatic variables. This exercise confirmed that some Great Soil Groups – red soils, desert soils, reddish-brown soils, and reddish-chestnut soils, for instance – do have limits set by climatic factors. There is some overlapping of Great Soil Groups owing to other influences on soil distribution, but a broad relationship between soil type and climate in the western United States emerges from the analysis.

Arkley's study led to two important conclusions. First, the amount of water available for leaching has a large influence upon soil genesis. Soils with low base saturation or strong acid reaction (or both) form under climates in which leaching effectiveness is high (more than 46 cm in warm conditions and 30 cm in cold conditions), and in environments where local conditions such as coarse texture enhance leaching. Examples are podzols, brown podzolic soils, grey-brown podzolic soils, *sols bruns acides*, and reddish-brown lateritic soils. Soils generally leached of carbonates and retaining a quantity of exchangeable bases within the solum are fostered under climates with moderate leaching (a leaching effectiveness of 15 to 40 cm). Examples are prairie soils, non-calcic brown soils, brown forest soils, and grey wooded soils. Soils tending to retain bases in all, or a large part of, the solum evolve under climates where leaching is slight (leaching effectiveness less than 15 cm) and no water can be expected to penetrate below the rooting depth of ordinary plants. Examples are red desert soils, desert soils, sierozems, reddish-chestnut soils, chestnut soils, chernozems, and parts of the western brown forest and grey wooded soils. Second, soil genesis is influenced by rates of actual evapotranspiration. In warm and hot climates with low actual evapotranspiration, soils low in organic matter are formed – red desert soils, desert soils, and sierozems. In cooler climates with moderate evapotranspiration, soils low in organic matter also form – reddish-brown soils, non-calcic brown soils, and brown soils. Under climates where actual evapotranspiration is high, all soils tend to be rich in organic matter, save for some reddish-brown lateritic soils where high leaching, intense organic matter decomposition, and low fertility lead to low organic matter content.

Arkley's work suggests that important soil processes are influenced by the Earth's surface energy and water balances, and that, because of this, there is a broad relationship between soil types and climate. Many pedologists would claim that soil–climate relationships are close enough to predict the zonal soil type from a knowledge of climate, all other factors (organisms, topography, parent material, and time) being constant. A zonal soil can be regarded as a soil in a steady state – that is, a soil adjusted to prevailing environmental, and especially climatic, conditions. Were climate and all other environmental factors known to be stable over time-scales of soil formation, then there would be justification for hypothesizing that distinct soils types would be nurtured by different climatic regimes. It has become abundantly clear over the last two decades that climate and other environmental factors are *not* stable, but constantly change in response to internal system thresholds and external forcing. The implications of this environmental dynamism for soil–climate relationships established within a developmental framework are serious. Climate has changed significantly in most parts of the world, even during the relatively short Holocene epoch. These climatic changes are certain to have influenced soil genesis. So how much credence should be given to relationships between soil types and climatic parameters derived from data collected over the last century? To answer this, it is necessary to know if regional climatic variations in the Holocene epoch were greater than present-day regional climatic variability. For many regions a considerable shift of the climatic zones has occurred during the Holocene, and climates have changed significantly. Of course, it may be that the zonal soils represent a response to the average climatic conditions over the last 10,000 years. These issues are difficult to resolve but, until they are, any suggestion that present zonal soil types are in equilibrium with climate is questionable (cf. Ruhe 1983).

Although it is disputable whether zonal soils are really climax systems, there is plenty of evidence suggesting that some systems of the pedosphere are in a climatically determined steady state. Soil systems with rapid turnover times, such as litter horizons, may well respond quickly to climate and so track climatic change rapidly and maintain a steady state that is always approximately in balance with the prevailing climatic regime. It was shown in the previous chapter that several soil properties (texture, field capacity, litter production, decomposition rates of wood and litter, acidity, and base saturation) follow the global throughput pattern that mirrors the simultaneous availability of heat and moisture. Other soil properties (litter accumulation and carbon content) adhere to an accumulation pattern that reflects gains of material over long time-spans. This raises the question of scale-effects on pedogenesis (e.g. Yaalon 1971), a little explored avenue of enquiry: far more work is required on soil evolution at different spatial and temporal scales.

SOIL CLIMOSEQUENCES

The old idea of soil zonality had different climatic zones engendering distinct soil types with very sharp pedotones (the pedological equivalent of an ecotone) between them. In contrast, the ecological and evolutionary view of geoecosystems promulgated in this book lays emphasis on climatic fields or gradients between core climatic regions. The response of soil properties to climatic gradients may be investigated empirically using the 'clorpt' formula or 'brash' formula. A soil property is measured along a climatic gradient at sample points chosen carefully to minimize the influence of other environmental factors. The resulting relationship, which may be plotted as a graph and described by a suitable regression equation, is known as a climofunction and the changing soil property a climosequence. Most climosequences relate climatically sensitive soil properties to mean annual temperature or mean annual rainfall, or to some measure of effective rainfall. They are established mainly for soils within macroscale landscapes, for at smaller scales it becomes difficult to winnow the climatic effects from topographic and historical effects. Where the climatic gradients are a response to altitude, much more local differences in soil properties can be attributed to climate (see Chapter 5).

Organic components

Soil organic matter is a major component of most geoecosystems. The amount of organic matter in soil represents a balance between production of plants and decomposition of organic detritus. Influences on these processes are complex but climate appears to play a leading role in macroscale and megascale landscapes. Carbon and nitrogen are primary ingredients of soil organic matter, and climosequences for these chemicals have been established in a variety of regions. A pioneering study was carried out by Jenny (1941: 171), who investigated the influence of rainfall and temperature on the total organic nitrogen content of loamy surface soils (0 to 20 cm) of former grasslands in the North American Great Plains. For a fixed temperature, soil nitrogen content increases along gradients of increasing humidity from south to north. The rate of increase is steeper at lower mean annual temperatures,

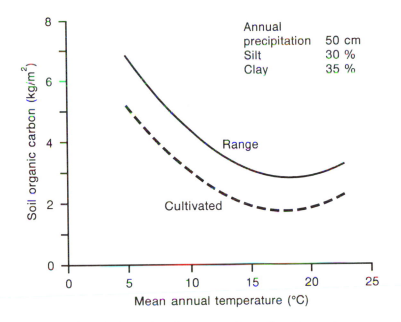

Figure 3.1 Predicted relationships between soil organic carbon and mean annual temperature, with mean annual precipitation fixed at 50 cm, in range and cultivated grassland soils in the United States
Source: After Burke *et al.* (1989)

and the soil nitrogen content of soils in the northern part of the Great Plains is greater than the nitrogen content in the southern part. The increase is exponential: nitrogen contents rise faster per unit increase in moisture as the Canadian border is approached. For a fixed moisture level, soil nitrogen content decreases from north to south. The decline is exponential, the nitrogen level falling a bit faster in northern regions than in southern regions.

Predictions made using single environmental factors are of limited worth because controls on soil organic matter are many and complex. A particularly interesting study tried to predict soil, nitrogen, and phosphorus dynamics in the central grassland region of the United States using simultaneously changing controls (Parton *et al.* 1988; see also Parton *et al.* 1987). A later study sought to establish quantitative relationships between soil organic matter levels in central plains grasslands and key driving variables: precipitation, temperature, and soil texture (Burke *et al.* 1989). Soil properties (organic carbon, organic nitrogen, sand, silt, clay, and bulk density for the top 20 cm) were calculated from raw data for about 500 pedons in rangeland soils and some 300 pedons in cultivated soils. Climatic data (mean annual temperature or growing season temperature, mean annual precipitation or

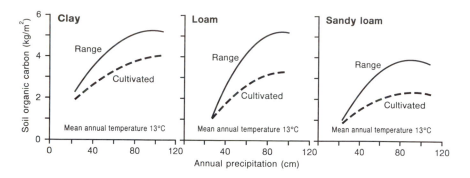

Figure 3.2 Predicted relationships between soil organic carbon and annual precipitation, with mean annual temperature fixed at 13 °C, in range and cultivated grassland soils in the United States. Three different soil textures are shown: (a) clay (50% clay, 20% silt), (b) loam (20% clay, 40% silt), and (c) sandy loam (10% clay, 30% silt)
Source: After Burke *et al.* (1989)

growing season precipitation) were culled from records held by the US Weather Bureau at some 600 sites. Regression analysis was performed on all possible data subsets to find the best predictive equation for soil organic carbon and nitrogen, rangeland and cultivated data sets being analysed independently. In all cases, the best temperature predictor was mean annual temperature, the best precipitation predictor was mean annual precipitation, and the best texture predictors were silt and clay taken separately. Finally, mean annual temperature, mean annual precipitation, silt, and clay were entered into a full quadratic model. This was then used to predict the mass of soil organic carbon in the top 20 cm of soil. It was found that soil organic carbon in rangeland and cultivated soils decreased with mean annual temperature to about 17°C (Figure 3.1), a trend attributable to increasing decomposition rates. The slight rise in soil organic carbon at temperatures above 18 °C was probably an artefact of the quadratic regression model. Soil organic carbon in range and cultivated soils responded strongly to mean annual precipitation. It rose to an annual precipitation mean of about 80 cm, then levelled out, and then fell a little at an annual precipitation mean of 100 cm, though this slight fall may have been an artefact of the quadratic regression model (Figure 3.2). The trends in organic carbon with increasing precipitation may be explained by increasing plant productivity and consequently carbon inputs to the soil. The levelling out at a mean annual precipitation of 80 cm may be interpreted as the net effect of decomposition rates increasing as rapidly as production rates with more than that amount of precipitation. Combined, the predicted relationship of soil organic carbon to the two climatic variables is shown in Figure 3.3. Soil organic nitrogen contents in the samples were so closely correlated to soil organic carbon that

54

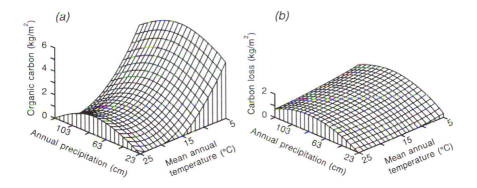

Figure 3.3 (a) Soil organic carbon 'surface' predicted from mean annual temperature and mean annual precipitation on loam soils (20% clay, 40% silt). (b) Predicted carbon loss owing to cultivation
Source: After Burke *et al.* (1989)

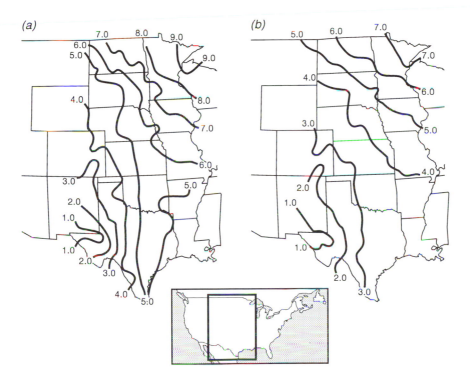

Figure 3.4 Regional patterns of soil organic carbon (kg/m²) in loam soils (20% clay, 40% silt). (a) Range soils. (b) Cultivated soils
Source: After Burke *et al.* (1989)

55

Table 3.1 Soil climosequences on greywacke in New Zealand

Soil number	Mean annual precipitation (cm)	Nitrogen in top 20 cm (%)	Entire pedon to bedrock		
			Carbon (kg/m²)	C/N ratio	Total phosphorus (kg/m²)
1	31–38	0.111	4.04	8.8	0.313
2	38–46	0.163	6.39	8.5	0.349
3	51	0.279	10.24	13.1	0.259
4	140–152	0.242	11.21	16.9	0.313
5	152–203	0.242	13.79	18.9	0.191
6	381	0.219	11.96	21.5	0.126

Source: After Walker and Adams (1959)

the carbon data, in conjunction with clay content, could be used to predict nitrogen. The regression equation for soil carbon was used to predict soil organic carbon across the central plains grassland using an extensive US Weather Bureau database and holding soil texture constant as loam (20 per cent clay, 40 per cent silt, and 40 per cent sand). Results show that soil organic carbon should generally increase eastwards across the Great Plains as annual precipitation rises, with lowest values in the south-west and highest values in the north-east (Figure 3.4).

Soil carbon and nitrogen climofunctions have also been established for soils formed in greywacke sandstone in New Zealand (T. W. Walker and Adams 1959). The soils were sampled along the 8°C mean annual temperature isotherm, on slopes of between 9 and 27 per cent, under tussock grassland and subalpine scrub tussock. Soil nitrogen content, carbon content, carbon–nitrogen ratio, and total phosphorus content, all appear to vary with increasing precipitation (Table 3.1). Notice that the nitrogen content in the top 20 cm peaks at a mean annual rainfall of about 50 cm, whereas the carbon content in the entire pedon seems to peak at about 175 cm.

Inorganic components

Other relatively mobile constituents of soils are influenced by climate. Mobile forms of aluminium, iron, and phosphorus are subject to climatic influence. Peter W. Birkeland and his colleagues (1989) calculated accumulation indices for pedogenetically significant aluminium and iron, and a depletion index for phosphorus, in soil chronosequences from Baffin Island in the Canadian Arctic, the alpine Sierra Nevada and Wind River Range in the western United States, the alpine Khumbu Glacier area of Mount Everest in Nepal, and the alpine Southern Alps in New Zealand. When ranked according to the degree of soil development, these indices provided a sequence related to climate. The greatest accumulation and depletion had occurred in the warmest and wettest environment, and the least in the coldest and driest environment.

Leaching of calcium carbonate is particularly sensitive to climate. One of the most credible climofunctions relates the depth of carbonate accumulation (the lime horizon) to mean annual rainfall along a transect running from the semiarid parts of Colorado, through Kansas, to the humid areas of Missouri (Jenny and Leonard 1934). The soils along the transect, which follows the 11 °C isotherm, are all formed in loess and lie on broad ridges. The regression equation describing the climofunction is

$$d = 2.5 \, (P - 12)$$

where d is depth to carbonates (cm) and P is mean annual precipitation (cm). Arkley (1963) established a similar climofunction for soils developed on old mesas, alluvial terraces, and fans in arid Nevada and California. The regression equation for these soils was

$$d = 1.63 \, (P - 0.45)$$

Notice that the regression coefficients determining the slope of the regression line (and thus the rate of increase in depth-to-carbonates with increasing rainfall) are smaller in the arid West than in the Great Plains. This suggests that the cold winter rains falling in California are more effective displacers of calcium carbonate than are the warm summer showers in Colorado, Kansas, and Missouri (Jenny 1980: 326). In desert soils of the North American Southwest, Giles M. Marion (1989) correlated the long-term calcium-carbonate-accumulation rate, Ca (g/m^2/yr), with present-day mean annual precipitation, P (mm). The relationship was

$$Ca = 0.015 \, (P - 37)$$

Conclusions and caveat

The results of climosequence analysis for regions with a mean annual rainfall of 380 to 890 mm may be summarized as follows: increasing mean annual rainfall is associated with decreasing acidity, increasing depth to carbonates, increasing nitrogen content, and increasing clay content in the solum; while increasing temperature leads to decreasing contents of organic matter and nitrogen, increasing clay content, and redder colours. These soil properties would be expected to reflect the thermal and moisture patterns of macroscale and megascale landscapes.

A word of caution is advisable at this juncture: relatively few rigorously evaluated and reliable climosequences have been established since Jenny's pioneering work in the 1930s (Yaalon 1975; Huggett 1982; Catt 1988). The chief reason for this seems to lie in the difficulties of gauging the effect of climate on a soil property whilst holding all the other environmental factors constant, for the reasons mentioned in the discussion of the 'brash' equation (Chapter 2). It is usually the case that many environmental factors are not truly independent variables, that some factors are not amenable to quantification, and that the constancy of most of the factors is very difficult to establish. These difficulties are illustrated by the case of the carbonate–climate function (Catt 1988: 541). A prerequisite to establishing the relation between depth-to-carbonates and mean annual rainfall is that a number of sites can be located where the following conditions are fulfilled: rainfall has been consistently different but other climatic factors have been the same; and the soils at each site have been evolving for the same length of time, in very similar parent materials, in similar geomorphological situations, and have all experienced the same changes of vegetation and the same disturbance by animals. Some of these prerequisites, such as geomorphological situation and vegetational history, can be met by field observations and laboratory work on the pollen content of the soils or, failing that, of nearby lakes. The similarity of faunal influence at all sites may, perhaps, reasonably be assumed. Other preconditions are more evasive. Save for exceptional cases, such as soils formed on volcanic deposits, the age of a soil has to be assessed indirectly using stratigraphical evidence or pedological evidence. Both these lines of evidence for soil age are beset with problems. Decalcification is a slow process, its effects being measurable over thousands of years or much longer. Over time intervals of that duration, annual rainfall is hardly likely to have held constant. If all sites had experienced the same climatic changes, this inconstancy of annual rainfall rate might not be too serious a drawback, but normally the sites will be in different regions, each of which will have had a different climatic history.

Some of the problems discussed by John Catt (1988) are illustrated by Robert V. Ruhe's (1984) demolition of a classic climosequence in the midwestern United States. The climatic gradient believed to cause this climosequence runs across the prairies, from eastern Iowa to western Kansas. Ruhe disputed the validity of this climosequence because the geoecosystem in which the soils have evolved has itself evolved in a manner far more complex than simple soil–climate relationships would have us believe. The soils in the sequence are all Mollisols. They are divided at about longitude 96° to 97° W into Udolls, lying to the east, and Ustolls, lying to the west. The relationships between soils and climate are complicated by soil stratigraphy and climatal and vegetational change. The Udolls are formed in Peoria loess of late Pleistocene age and the maximum duration of weathering in them, as established by radiocarbon dating, is 14,000 years. This contrasts with the Udolls, formed in Bignell loess of Holocene age, where weathering can have occurred for a maximum of 9,000 years. The 5,000-year head start by the Udolls directly affected the climatic impact on the soils. Before the Ustolls originated, the Udolls supported coniferous and then deciduous forests under a moist climate. The deposition of the Bignell loess in which the Ustolls evolved appears to have taken place under a severe climate with hot, arid summers followed by cold, dry winters. For 5,000 years, the Udolls and Ustolls evolved under a climate some 40 to 50 per cent drier than today, with moisture deficits increasing in magnitude and annual duration from east to west. The weathering record in the soils is chiefly contained in their base status, which varies inversely with age and climate. The base status of the Udolls is lower than that of the Ustolls and varies less with longitude. This is probably the result of 5,000 years' extra weathering and two additional spells of water surplus. The base status of the Ustolls is higher than that of the Udolls and varies more with longitude. Indeed, the base status changes across the Udoll soil sequence may be a true climofunction developed during the Holocene epoch. But the major change in the loessal sediments and the chemical and physical properties of soils formed in them is a discordance at about longitude 96° to 97° W at the junction of the Peoria and Bignell loesses. This discordance is not climatically determined.

Considering these problems, it is not surprising that trustworthy climofunctions have been unforthcoming except in a few cases, such as soil nitrogen, where the soil property has a short turnover time and rapidly attains a steady state. Thus, it is possible to use climatic variables to predict the size of the litter store at a global scale (e.g. Esser and Lieth 1989). Clearly, the evolution of entire geoecosystems is usually too complex for univariate relationships to be extracted. This is patently the case in the climosequence across the American prairies, where the present soils are the outcome of many environmental influences, the relative importance of which have changed with time. The suggestion is not that climate fails to influence soil properties; but that, given the dynamism of the environment, climatic effects will be masked by the effects of other environmental influences and very hard to decipher. A dynamic systems model on the general lines of the 'brash' equation might help explain some effects, but no such model has yet been built. In the meantime, the best course might be to use the multivariate statistical techniques that have proved so successful in ecology (as will be seen in the next chapter) to establish the basic dimensions of the problem.

SUMMARY

The relationships between soils and other environmental factors vary with the scale of study. At megascales and mesoscales, climatic factors are important determinants of soil types and soil properties. The developmental view of zonal soils stresses the overriding role of climate in determining the 'formation' of the chief soil types. However, seen in an evolutionary context, where environmental dynamism is the keynote, the notion of soils developing towards some mature or 'climax' form appears suspect. Climatic fields appear to engender regular variations in soil types or properties known as climosequences. Much of the work on climosequences considers the relationship between a single soil property and one, or at most two, climatic variables. The truly ecological view incorporated in the 'brash' equation demands that the multivariate nature of soil relationships within the geoecosphere be studied. Likewise, the evolutionary aspect of the 'brash' equation, stressing an environmental dynamism in which 'soil-forming factors' themselves change with time, indicates that univariate or bivariate climofunctions are a gross simplification of the relationships between soils and other components of geoecosystems, except perhaps for soil properties, such as carbon and nitrogen levels, that respond relatively rapidly to changing climatic conditions. A new look at the classic climosequence across the North American prairies supports this contention.

FURTHER READING

The relations between climate and soils are discussed in many textbooks concerned with soils and related topics. For those wishing to acquaint themselves more fully with the concept of zonal soils, a good, short, and readable introduction is E. Mike Bridges' *World Soils* (1978). *Soil Genesis and Classification* (Buol *et al.* 1980) is more detailed, but highly informative about the USDA soil classification. References to soil–climate relationships are less common. Jenny's *The Soil Resource* (1980) has a chapter devoted to soil–climate relationships, and particularly climosequences, and the book by Birkeland entitled *Soils and Geomorphology* (1984) has some useful information. A discussion of soils and climate may also be found in *Climate, Earth Processes and Earth History* (Huggett 1991). The paper by Parton and his colleagues (1987) shows how a dynamics systems model can be used to predict regional variations in the state of a soil system component.

4

CLIMATE AND LIFE

Organisms are a vital component of geoecosystems. Individually, even the largest of them occupy very little space. As a whole, they form an almost continuous film over the land surface and in the edaphosphere. Single organisms are basic units in the genealogical and societary hierarchies. In the genealogical hierarchy, they are storehouses of genetic information – genomes; in the societary hierarchy, they are members of local populations. From an evolutionary point of view, these twofold aspects of organisms should be treated together (Eldredge 1985). Changes in the genealogical hierarchy supply the means by which organisms, expressed as phenotypes, can adapt over generations to evolving biotic and abiotic environmental circumstances, including changing climatic fields. As members of populations, individual organisms are part of societies or communities that, over time, evolve to suit the environmental fields in which they are engulfed, providing that these fields stay constant long enough for adaptation to occur. In this chapter, the ways in which the present climatic fields influence the distribution of species and communities will be explored.

THE DISTRIBUTION OF SPECIES

A major thrust of research in physiological ecology has been to elucidate the effect of environmental factors on the survival, growth, and reproduction of organisms. Organisms live in virtually all landscapes, from the hottest to the coldest, the wettest to the driest. Only in small areas that are intensely heated by volcanic activity do high temperatures preclude life. Bacteria can grow in the superheated water of geysers and deep-sea vents, and produce films in the boiling water of hot springs. At the other extreme, lichens can still photosynthesize at −30 °C, providing they are not covered with snow. The reddish-coloured snow alga, *Chlamydomonas nivalis*, lives on ice and snow fields in the polar and nival zones, imparting a pink look to the landscape. Many organisms are adapted to life in water. Aridity poses a problem of survival, but species of algae have been found in the exceedingly dry Gobi desert. Higher plants survive in arid conditions by xerophytic adaptations

61

that enable them to retain enough water to keep their protoplasts wet.

A vast range of environmental conditions lies between the extremes found in the Earth's landscapes. An individual species can live only within a certain range of environmental factors – the effect of too much or too little of any one factor may inhibit growth or even prove fatal. An environmental factor that retards or inhibits the growth of a species is a limiting factor. This term was suggested by Justus von Liebig (1840), a German agricultural chemist. Liebig observed that the growth of a field crop is hampered by whatever nutrient happens to be in short supply, and proposed a 'law of the minimum': the productivity, growth, and reproduction of organisms will be constrained if one or more environmental factors lie below their limiting levels. It was later observed by ecologists that there is also a 'law of the maximum' that applies where an environmental factor constrains organisms above a limiting level. So, for each environmental factor there is a lower below which the species cannot grow at all, an optimum at which the species thrives, and an upper limit above which no growth occurs (Blackman 1905). The upper and lower bounds define the tolerance range of the species for a particular environmental factor. A population will thrive within the optimum range of tolerance; survive but show signs of physiological stress near the tolerance limits; and not survive outside the tolerance range (Shelford 1911).

Much of the modern literature dealing with the interaction of organisms and their environment emphasizes the notion of stress. A polemical term, stress may be taken as 'external constraints limiting the rates of resource acquisition, growth or reproduction of organisms' (Grime 1989). As the conditions near the margins of tolerance create stress, it follows that the geographical range of a species is strongly influenced by its ecological tolerances. It is generally true that species with wide ecological tolerances are the most widely distributed. A species will occupy a habitat that meets its tolerance requirements, for it simply could not survive elsewhere. None the less, even where a population is large and healthy, not all favourable habitat inside its geographical range will necessarily be occupied, and there may be areas outside its geographical range where it could live. To an extent, the actual range of a species is a dynamic, statistical phenomenon that is constrained by the environment: in an unchanging habitat, the geographical range of a species can shift owing to the changing balance between local extinction and local invasion. And, it may also enlarge or contract owing to historical factors; witness the spread of many introduced species and chance colonizers in new, but environmentally friendly, regions – the spread of the American muskrat (*Ondatra zibethicus*) in central Europe after the introduction of five individuals by a landowner in Bohemia in 1905 (Elton 1958), and the establishment of the ladybird *Chilocorus nigritus* in several Pacific islands, north-east Brazil, West Africa, and Oman after shipment from other areas (Samways 1989) are examples.

Fine-scale studies

Owing to complexities involving life history, acclimatization, escape mechanisms, biotic interactions, and multivariate responses, tolerance ranges of
species are difficult to measure in the field. The tolerance range of a particular
species will depend on the stage of its life cycle. In plants, for instance, the
tolerance range will differ at the following stages: seeds, germination, growth,
and flowering. The lethal temperatures for some species depend on the
temperature at which they have been living, in other words, on how well they
have acclimatized to the environment. Many animals are able to escape severe
conditions – migratory birds and insects avoid polar winters; others
hibernate or aestivate to escape times of environmental harshness. Populations of organisms interact among themselves through mutualism, parasit-
ism, predation, and so forth; this biotic interaction may involve limiting
factors of a biotic variety. A key issue is that tolerances to different
environmental factors are not independent. To belabour the point, it is not
possible to take a component equation from the 'brash' equation and let one
factor change while all others are held constant – the factors are not
independent. For instance, in many fish, temperate tolerance will be
significantly influenced by acidity. Despite these difficulties, tolerance ranges
have been established for some species and used to explain their geographical
limits. We shall illustrate this by briefly looking at climatic constraints on
plant species' distributions.

Many distributional boundaries of plant species seem to result from
extreme climatic events causing the failure of one stage of the life cycle (Grace
1987). The climatic events in question may occur rarely, say once or twice a
century, and the chances of observing a failure are slim. None the less, edges
of the geographical ranges of plants often coincide with isolines of climatic
variables. Edward James Salisbury (1926) established that the northern limit
of madder (*Rubia peregrina*) in northern Europe sits on the 40°F mean
January isotherm. Johannes Iversen (1944), by carefully studying holly (*Ilex
aquifolium*) during a run of very cold winters (1939 to 1942), noticed severe
frost damage. He showed that the species was confined to areas where the
mean annual temperature of the coldest month exceeds –0.5 °C, and, like
madder, seemed unable to withstand low temperatures. Several frost-
sensitive plant species, including *Erica erigena*, *Daboecia cantabrica*, *Pinguicula grandiflora*, and *Juncus acutus*, occur only in the extreme west of the
British Isles where winter temperatures are highest. Other species, such as
Linnaea borealis and *Trientalis europaea*, have a northern or north-eastern
distribution, possibly because they have a winter chilling requirement for
germination that southerly latitudes cannot provide (Perring and Walters
1962). Low summer temperatures seem to restrict the distribution of such
species as the stemless thistle (*Cirsium acaule*). Near to its northern limit, this
plant is found mainly on south-facing slopes, for on north-facing slopes it

fails to set seed (Pigott 1974). The small-leaved lime (*Tilia cordata*) does not set seed at the northern limit of its distribution unless the summer is particularly warm (Pigott and Huntley 1981). Similarly, the distribution of grey hair-grass (*Corynephorus canescens*) is limited by the 15 °C mean isotherm for July. This may be because its short life span (2 to 6 years) means that, to maintain a population, seed production and germination must continue unhampered (Marshall 1978). At the northern limit of grey hair grass, summer temperatures are low, which delays flowering, and, by the time seeds are produced, shade temperatures are low enough to retard germination.

There is a danger of being lured into false conclusions when comparing plant distributions with climatic maps because a single factor – a climatic variable – is extracted from a rich and intricate web of biotic and abiotic interactions. Species at the limits of their distribution will be affected by many factors other than climate including soil, topography, microclimate, and competition with other species. Sorting out these multivariate inter-actions by implementing a series of experiments and a field programme is costly and consumes time (Grace 1987), but it should improve understanding of climatic influences on plant species distribution. Experiments and field-work have been conducted on a range of individual species. To take but one example, foliar frost resistance in Australian temperate and tropical forest tree species has been studied by Jennifer Read and her colleagues (Read and Hill 1988, 1989; Read and Hope 1989). This work has generated several interesting ideas about the geographical distribution of individual tree species. It shows that frost resistance generally accords with the known history and geographical and climatic ranges of species, but brings to light some exceptions. For instance, the frost resistance of the southernmost species of southern beech, *Nothofagus cunninghamii*, has a higher foliar resistance than its more northerly relatives. On the other hand, *Phyllocladus aspeniifolius* is more common at high altitudes than *Eucryphia lucida*, but has a lower foliar frost-resistance, and *Athrotaxis selaginoides* does not have the superior frost resistance that one would expect, given that it occurs at higher altitudes than *Nothofagus cunninghamii*.

Models have also helped to probe the relations between climatic factors and species' distributions. A climatic model has been used to predict the distribution of woody plant species in Florida, USA (Box *et al.* 1993). The State of Florida is small enough for variations in substrate to play a major role in determining what grows where. None the less, the model predicted that climatic factors, particularly winter temperatures, exert a powerful influence, and in some cases a direct control, on species' distributions. James M. Lenihan (1993) investigated the climatic response of boreal tree species in North America. He used several climatic predictor variables in a regression model. The variables were annual snowfall, degree-days, absolute minimum temperature, annual soil–moisture deficit, and actual evapotranspiration summed over summer months. Predicted patterns of species' dominance probability closely matched observed patterns. This suggested that the dominance probability of particular species represents an individualistic response to different combinations of climatic constraints across the region. Other studies support the finding that plant taxa respond on an individual basis to different climatic variables (e.g. Huntley *et al.* 1989).

Broad-scale studies

Studies on individual species are useful but a problem arises: how may the results from several different studies be drawn together to make useful generalizations and predictions about the organization of communities? Without some attempt at broad-scale synthesis, it is not easy to distinguish between the response of a single species and a general biogeographical process (cf. J. H. Brown and Maurer 1989). Although high resolution, autecological studies have their value, much can be learnt about environmental limits by studying species at a lower resolution. Studies on a broad scale have been made more feasible of late by the development of multivariate statistical techniques, and by the compilation of detailed distribution maps of species. A top-notch example of this work analyses the distribution patterns of *Salix* species (willows) in Europe, as performed by Åse Myklestad and H. John B. Birks (1993). There are sixty-five native species of *Salix* in Europe. The occurrences of these in 484, $2° × 2°$ (latitude × longitude) grid squares comprising Europe were recorded from distributional maps in the *Atlas Florae Europaeae* (Jalas and Suominen 1976). For each recording area, the values of twelve environmental, one historical, and twelve morphological variables were recorded. These data were obtained for all bar four species, so giving a matrix of sixty-one species and twenty-nine habitat and morphological variables. The matrix was subjected to methods based on the weighted-averaging ordination technique called correspondence analysis. This technique tries to pick out major patterns in the data as the first few ordination axes and to demote individual responses and noise to lower-order axes. The patterns so discerned can be described as classificatory groups by means of a two-way indicator species analysis (TWINSPAN) or as gradients by means of, for example, detrended correspondence analysis (DCA). Ordination axes commonly seem to reflect underlying environmental gradients (see R. H. Whittaker 1967). Myklestad and Birks (1993: 2) rightly caution that as 'the human mind can often think up explanations for almost any pattern, even non-significant ones, the proposed relationships of the observed patterns in the species data should be rigorously tested'. A suitable test is canonical correspondence analysis (CCA) which determines the extent to which species are related to environmental variables and the relative importance of the environmental variables in explaining species' patterns.

Eigenvalues of the first four axes produced by DCA ($\lambda_1 = 0.51$, $\lambda_2 = 0.21$,

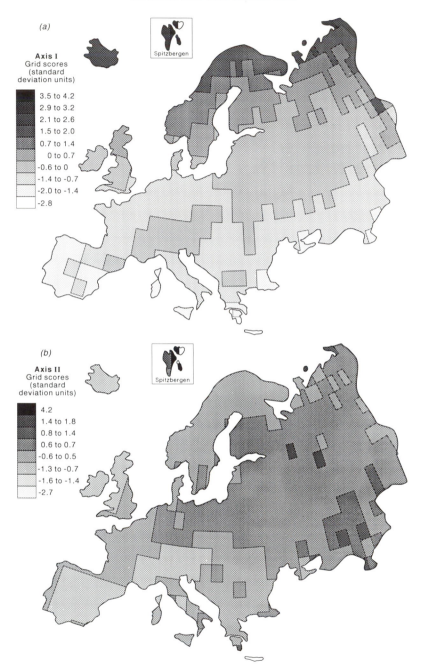

Figure 4.1 Regional patterns of *Salix* assemblages as revealed by DCA. (a) DCA axis
I. (b) DCA axis II. Grid scores are expressed in standard deviation units
Source: After Myklestad and Birks (1993)

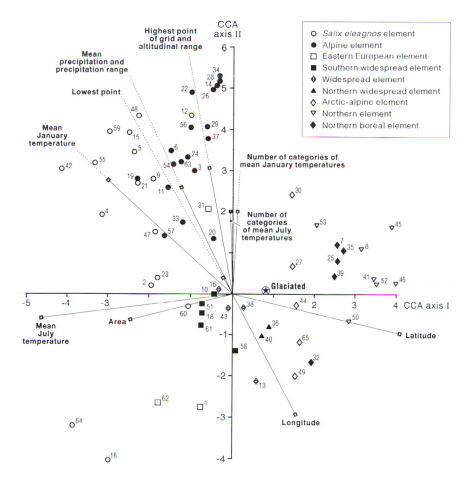

Figure 4.2 Sixty-five *Salix* species and thirteen environmental variables for Europe plotted within axes I and II of a CCA ordination. Note that the environmental variable referred to as 'glaciated' is a nominal variable. The scale marks are in standard deviation units. The scores for the continuous environmental variables have been multiplied by ten. The TWINSPAN group to which each species belongs is indicated. The biplot captures 26.4 per cent of the total biological variation. The species, identified by numbers, are: 1 *Salix acutifolia*. 2 *Salix alba*. 3 *Salix alpina*. 4 *Salix amplexicaulis*. 5 *Salix appennina*. 6 *Salix appendiculata*. 7 *Salix arbuscula*. 8 *Salix arctica*. 9 *Salix atrocinerea*. 10 *Salix aurita*. 11 *Salix bicolor*. 12 *Salix breviserrata*. 13 *Salix burjatica*. 14 *Salix caesia*. 15 *Salix cantabrica*. 16 *Salix caprea*. 17 *Salix caspica*. 18 *Salix cinerea*. 19 *Salix crataegifolia*. 20 *Salix daphnoides*. 21 *Salix eleagnos*. 22 *Salix foetida*. 23 *Salix fragilis*. 24 *Salix glabra*. 25 *Salix glauca*. 26 *Salix glaucosericea*. 27 *Salix hastata*. 28 *Salix hegetschweileri*. 29 *Salix helvetica*. 30 *Salix herbacea*. 31 *Salix hibernica*. 32 *Salix jenisseensis*. 33 *Salix kitaibeliana*. 34 *Salix laggeri*. 35 *Salix lanata*. 36 *Salix lapponum*. 37 *Salix mielichhoferi*. 38 *Salix myrsinifolia*. 39 *Salix myrsinites*. 40 *Salix myrtilloides*. 41 *Salix nummularia*.

Figure 4.2 (continued) 42 *Salix pedicellata.* 43 *Salix pentandra.* 44 *Salix phylicifolia.* 45 *Salix polaris.* 46 *Salix pulcra.* 47 *Salix purpurea.* 48 *Salix pyrenaica.* 49 *Salix pyrolifolia.* 50 *Salix recurvigemmis.* 51 *Salix repens.* 52 *Salix reptans.* 53 *Salix reticulata.* 54 *Salix retusa.* 55 *Salix salviifolia.* 56 *Salix serpillifolia.* 57 *Salix silesiaca.* 58 *Salix starkeana.* 59 *Salix tarraconensis.* 60 *Salix triandra.* 61 *Salix viminalis.* 62 *Salix vinogradovii.* 63 *Salix waldsteiniana.* 64 *Salix wilhelmsiana.* 65 *Salix xerophila*
Source: After Myklestad and Birks (1993)

$\lambda_3 = 0.16$, $\lambda_4 = 0.11$) suggest that only the first two axes, which account for 28.4 per cent of the total biological variation, will express ecologically useful information. Figure 4.1 shows the geographical pattern of change in *Salix* assemblages along the first and second DCA axes. The trend along the first axis displays latitudinal change in species composition from Svalbard, with extreme positive scores, to the Mediterranean area, with extreme negative scores. Mountainous areas lead to a more northerly type of species composition extending southwards. In detail, the latitudinal bands are distorted a little and run from south-west to north-east, paralleling the turning of the isotherms of July mean temperatures. The trend on the second axis adds a further dimension to the pattern on the first axis. It shows a gradient running from dry lowland, mostly in continental areas of eastern and middle Europe, to mountainous areas of higher precipitation, and towards extremely oceanic areas in western Europe. To elucidate the environmental influences on these two trends, CCA was employed. As with DCA, the eigenvalues ($\lambda_1 = 0.45$, $\lambda_2 = 0.20$, $\lambda_3 = 0.10$, $\lambda_4 = 0.05$) suggest that the first two axes will capture most of the ecologically relevant information. The relative importance of the environmental variables is shown on the biplot (Figure 4.2). On the first axis, latitude, July mean temperature gradient, and possibly glaciation are important. On the second axis, the highest point in a grid-cell and altitudinal range are the most highly scoring variables, though area and mean annual precipitation appear to have some explanatory power. As would be expected, many of the environmental variables are intercorrelated (Figure 4.3). In an attempt to investigate more deeply the environmental effects on *Salix* distribution, a CCA was performed on the environmental variables with latitude and longitude excluded. The remaining nine environmental variables explained 26 per cent of the variance in the *Salix* data. Almost as much variance as this, 26 per cent, was explained by a cubic trend-surface describing the geographical co-ordinates of the grid-cells, pointing to a broadly similar spatial structure for the species and environmental factors that probably results from a similar response to a common underlying factor such as macroclimate. Some 49 per cent of the variance is unexplained, a high figure but not unexpectedly so. The causes of this variance are presumably independent of the eleven environmental variables included in the analysis, and are not well modelled by a cubic trend-

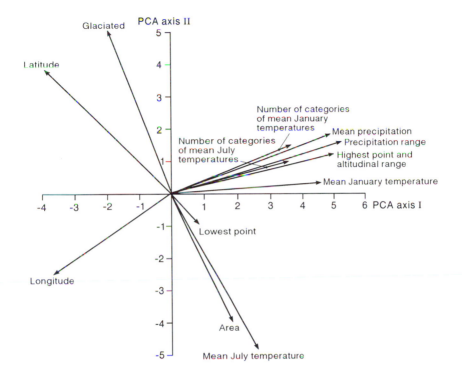

Figure 4.3 Relationships between thirteen environmental variables, as revealed by a PCA of the correlation matrix of the variables and the 484 European grid-cells, with only the thirteen variables shown here on the covariance biplot. Arrows pointing in the same direction indicate positively correlated variables; perpendicular arrows suggest a lack of correlation; arrows pointing in opposite directions indicate negatively correlated variables. Axis I accounts for 39.9 per cent of the total variation, axis II accounts for 20.3 per cent
Source: After Myklestad and Birks (1993)

surface of grid-cell latitudes and longitudes. Possibilities are a combination of local or regional abiotic, biotic, and historical factors; spatial processes more complex than a cubic trend-surface of latitude and longitude allows; and stochastic variation (cf. Borcard *et al.* 1992). Overall, the analysis strongly suggests that regional climate, mainly related to summer temperature, explains the distribution of *Salix* species in Europe. It was also found, again by performing CCA, that some types of species' distribution may relate to their occurrences in certain habitats and altitudes, possibly because of the temperature tolerances of those species (Figure 4.4). For reasons explained by Myklestad and Birks (1993: 21), the important variables on the first axis may be high-alpine habitat, maximum size, snowbed habitat,

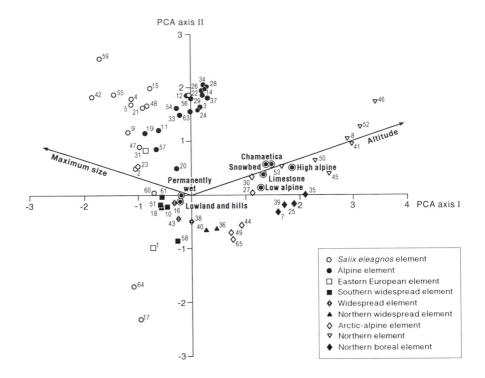

Figure 4.4 Twenty-nine habitat and morphological variables with an interset correlation >0.4 or <−0.4 (labelled in bold), and the sixty-five *Salix* species, plotted within axes I and II of a CCA ordination. The scale marks are in standard deviation units. The scores for the continuous environmental variables have been multiplied by ten. The TWINSPAN group to which each species belongs is indicated. The species, identified by numbers, are listed in the caption to Figure 4.2
Source: After Myklestad and Birks (1993)

limestone habitat, and permanently wet habitat. On the second axis, lowland-and-hill habitat is the only variable that may be associated with *Salix* species distribution.

To test possible relationships between *Salix* species richness and the environmental variables, multiple regression equations were computed with a forward progression of variables (Table 4.1). The importance of area and the variation in July temperature stand out as being important. When the effect of area was partialled out, the variance in the species richness explained falls to 31.9 per cent, which reinforces the well-known observation that richness of species is related to area. With area omitted, the two most important variables become variation in July temperature and latitude. So, *Salix* species richness is predicted to increase northwards, and in areas with higher

Table 4.1 Salix species richness versus environmental variables

Variable[a]	Correlation	t-value	Variance
All variables (multiple regression)[b]			
Area	0.45	11.1	0.2
Mean July temperature variability	0.39	6.6	0.09
Latitude	0.04	8.4	0.08
Lowest point	0.28	4.7	0.03
Mean January temperature variability	0.26	3.3	0.01
Mean precipitation	0.15	5.1	0.02
Precipitation range (highest value minus lowest value)	0.17	−3.1	0.01
Longitude	0.12	3.8	0.01
Area omitted (partial multiple regression)[b]			
Mean July temperature variability	0.33	6.6	0.09
Latitude	0.3	8.4	0.08
Mean January temperature variability	0.23	3.3	0.01
Mean precipitation	0.15	5.2	0.02
Lowest point (elevation)	0.19	4.7	0.03
Precipitation range (highest minus lowest)	0.12	−3.1	0.01
Longitude	0.06	3.8	0.02

Notes: [a]All variables refer to grid-cells
[b]With forward selection of variables; overall model significant at $p = 0.01$ in both cases
Source: After Myklestad and Birks (1993)

temperature variability, especially in July mean temperatures; it may also be influenced by the lowest point in a grid-cell and perhaps longitude and precipitation range. Interestingly, a similar multivariate analysis of *Salix* species in Fennoscandia shows that, as in the case of Europe, regional climate, and especially those components related to summer temperature, accounts for much of the species' distribution pattern (Myklestad 1993).

PLANT FORMATIONS

Animals and plants are part of a community, an assemblage of species sharing similar tolerances to environmental factors, including climate. Climate influences plant distributions through its effect on basic physiological processes. Plants and, to a lesser extent, animals possess structural and physiological adaptations that suit them for life in a particular habitat: they are adapted to light, heat, humidity, soil, terrain, and other organisms. These structural and physiological adaptations are reflected in life-form. The life-form of an organism is its shape or appearance, either its overall form (such as herb, shrub, or tree in the case of plants), as well as the form of its individual features (such as leaves). Animal life-forms are not much studied, largely because the major life-forms of animals match taxonomic categories, which is not usually the case with plants. Most mammals are adapted to basic habitats and may be classified accordingly: they may be adapted for life in water (aquatic or swimming mammals), underground (fossorial or burrowing mammals), on the ground (cursorial or running, and saltorial or leaping mammals), in trees (arboreal or climbing mammals), and in the air (aerial or flying mammals) (Osburn *et al.* 1903). None of these habitats is strongly related to climate. That is not to say that animal species are not adapted to climate: there are many well-known cases of adaptation to marginal environments (e.g. Cloudsley-Thompson 1975b).

Leaves

Many individual physiognomic features of vegetation are related to climate. For instance, the size and shape of leaves may be explained by climatic control. Leaves of plants of the same species differ according to whether they grow in the shade or in sunlight. Commonly, the leaves grown in the shade are larger and less dissected than leaves of specimens grown in direct sunlight. Similarly, plants tolerant of shady conditions in the understorey often possess larger and less lobed leaves than species growing in the sunny canopy. Remarkable is the independent evolution of similar leaf forms, especially of trees, under the same climate (Bailey and Sinnott 1916). In humid tropical lowlands, forest trees have evergreen leaves with no lobes. In regions of Mediterranean climate, plants have small, sclerophyllous evergreen leaves. In arid regions, stem succulents without leaves, such as cacti, and plants with entire leaf margins (especially among evergreens) have evolved. In cold wet climates, plants commonly possess notched or lobed leaf margins.

Comparative studies of terrestrial species of diverse lineages have disclosed clear ecological trends in some twenty-three aspects of leaf form, physiology, and pattern of arrangement, some of which are listed in Table 4.2. Many of these trends are established only qualitatively, and some may be influenced by phylogeny. An interesting interpretation of these trends considers the net contribution of the traits of a leaf, or a canopy, to carbon gain in an entire plant. And it explains the evolution of the traits as the outcome of energetic trade-offs between the economics of gas exchange, the economics of support, and the economics of biotic interactions (Givnish 1987).

Plant life-forms

A sophisticated and widely used classification of plant life-forms, based on the position of regenerating parts, was designed by Christen Raunkiaer (1934) who recognized five main groups: therophytes, cryptophytes, hemicryptophytes, chamaephytes, and phanerophytes. The diversity of these life-forms varies in different environments. Raunkiaer noticed that tropical forests contain a wide spectrum of life-forms, whereas in extreme climates,

Table 4.2 Some ecological trends in leaf form and function

Form or function	Trend
Effective leaf size (width of a leaf or its lobes or leaflets)	Increases along gradient of increasing rainfall, humidity, and/or soil fertility and decrease with increasing irradiance. Decreases with elevation on mountains in regions of high rainfall at low elevation and then decreases with elevation in more arid regions.
Leaf thickness	Increases with decreasing rainfall, humidity, and/or soil fertility. Increases with increasing irradiance and/or leaf lifetime. Tends to increase with elevation on mountains receiving high rainfall
Leaf absorptance in the visible spectrum	Tends to decrease (i.e. leaves become more glaucous or highly reflective) in sites that are sunnier, more arid, or less fertile
Leaf inclination from the horizontal	Tends to be greater in sunnier, more arid, or less fertile sites
Evergreen leaves	Common in habitats with nutrient-poor soils and/or little seasonal variation in favourability of conditions for photosynthesis (i.e. aseasonal or winter-rainfall climates). Plants with deciduous leaves predominate elsewhere, principally in deserts and semi-deserts, seasonal tropical forests, upper storeys of rain forests, and temperate forests of eastern Asia, eastern North America, and northern Europe
Leaves with non-entire margins (i.e. toothed or lobed)	Most common in dicots of north temperate zone and forest understoreys elsewhere
Lobed leaves	Common only in north temperate trees and in tropical trees of early succession
Leaves with long, acuminate drip tips	Common in wet rain forests and cloud forests, particularly among understorey species
Leaves with cordate (heart-shaped) bases	Common among vines, forest herbs, and aquatic herbs
Trees with compound leaves	Most common in arid and semiarid habitats that favour the deciduous habit, at low elevations, and in gap-phase succession

Source: After Givnish (1987)

with either cold or dry seasons, the spectrum is smaller (Figure 4.5). As a rule of thumb, very predictable, stable climates, such as humid tropical climates, support a wider variety of plant life-forms than do regions with inconstant climates, such as arid, Mediterranean, and alpine climates. Alpine regions, for instance, lack trees, the dominant life-form being dwarf shrubs

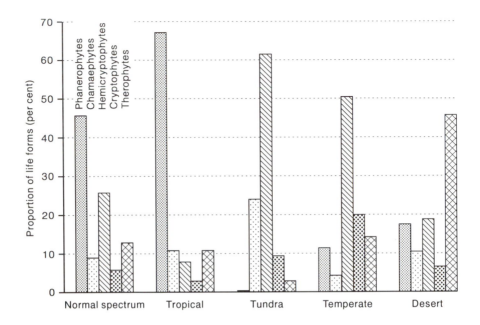

Figure 4.5 The proportion of life-forms in various environments. The 'normal'
spectrum is constructed by selecting a thousand species at random
Source: From data in Raunkiaer (1934) and Dansereau (1957)

(chamaephytes). In the Grampian Mountains, Scotland, 27 per cent of the
species are chamaephytes, a figure three times greater than the percentage of
chamaephytes in the world flora (Tansley 1939). Some life-forms appear to
be constrained by climatic factors. Megaphanerophytes (where the regenerating parts stand over 30 m from the ground) are found only where the
mean annual temperature of the warmest month is 10 °C or more. Trees are
confined to places where the mean summer temperature exceeds 10 °C, both
altitudinally and latitudinally. This uniform behaviour is somewhat surprising as different taxa are involved in different countries. Intriguingly, dwarf
shrubs, whose life-cycles are very similar to those of trees, always extend to
higher altitudes and latitudes than do trees (Grace 1987).

Given that a plant will only survive in an environment to which it is suited,
and that plant life-forms are to a large extent morphological adaptations to
environmental conditions, one would expect each different climatic type to
be associated with particular dominant life-forms. This appears to be the
case. The physiognomic vegetation types known as plant formations (or
biomes if animals be included) have long been recognized. Climate appears
to have shaped the evolution of structure and physiology of plants in each

formation-type (or zonobiome): plant species of very different stock growing in different areas, when subjected to the same climatic constraints, have evolved the same life-form. Tropical forests are characterized by climbing vines, epiphytes, and broad-leaved species. Very cold regions at high latitude or altitude are dominated by evergreen conifers. At even higher latitudes and altitudes, frost-resistant tundra species are the predominant life-form. The seasonal temperate zone supports broad-leaved deciduous trees where rainfall is moderate, tough-leaved evergreen shrubs where winter rains alternate with summer drought. Mediterranean vegetation occurs wherever a Mediterranean climate prevails, including Spain, Italy, Greece, southern California, Chile, the southern tip of Africa, and south-western Australia. In each of these regions, the life-forms of the plants are the same, but the species comprising the flora are different. The American cactus and the South African euphorbia, both living in arid regions, have adapted by evolving fleshy, succulent stems and by evolving spines instead of leaves to conserve precious moisture. These well-known relationships are but a few of the eye-catching correspondences between plant formations and climate. So striking are the correspondences that they caught the eye of Alexander von Humboldt and other naturalist-explorers. Humboldt (1817) believed that the floral zones occupying different altitudes and latitudes corresponded with the temperature zones encircling mountains and the globe.

The congruity between vegetation and climate is deemed close enough to be used as a basis for climatic classification, and vice versa. Two well-known classifications of vegetation and climate are due to Wladimir Peter Köppen (1931) and Leslie Rensselaer Holdridge (1947). Köppen's scheme was a classification of world climates based on the known distribution of vegetation types. The boundaries between different climates were chosen roughly to follow vegetation boundaries, and were expressed as aspects of climate, notably seasonality, that are relevant to plant growth. Holdridge's system was intended to predict natural potential vegetation from climate. It took two climatic variables. The first variable was biotemperature, defined as the sum of all mean monthly temperatures above $0\,^{\circ}C$. Biotemperature was assumed to relate directly to the potential amount of evaporation that would take place if unlimited soil moisture were available. The second variable was mean total annual precipitation. Using these variables as the axes of a diagram, Holdridge defined boundaries for altitudinal and latitudinal zones of the Earth and their corresponding vegetation units. In effect, he created climatic 'space' from two climatic variables, divided it by superimposing a regular grid, then allocated seemingly appropriate vegetation types to each grid-cell.

Climatic classifications such as Köppen's and Holdridge's assume that temperature and precipitation are the most effective climatic determinants of vegetation. Recent work has shown that this assumption is suspect, and that variables which integrate the energy and water balances, such as actual evapotranspiration, are more important determinants of plant growth. For this reason, genetic classifications of climate, which are based on the mechanisms that generate climate and which emphasize climatic core areas rather than climatic boundaries, may provide a better structure in which to frame investigations of climatic influences on vegetation. One such genetic schema, mentioned in Chapter 2, was devised by Heinrich Walter (1985). The transitional zones between the zonobiomes, which correspond to transitional zones between the chief genetic climatic types, are occupied by zonoecotones, belts where one type of vegetation yields to another, usually gradually, sometimes sharply (Figure 1.3, p. 19). An example is the transition zone created as deciduous forest gives way to temperate grassland. In belts of transition, two zonal vegetation types are found cheek by jowl. Competition is intense. The dominant vegetation type at a particular site depends on local conditions of climate, soil, and topography. Owing to the variation of local conditions, zonoecotones commonly display a mosaic pattern of the two vegetation types. Commonly, the climatic gradient across a zonoecotone is mirrored by changing proportions of the two vegetation types in the landscape mosaic.

An attempt to clarify the relation between zonobiomes and integrative climatic variables was made by Mikhail I. Budyko (1974: 364). In his analysis, Budyko used two climatic parameters. First, he used the radiative index of dryness, which characterizes the relative values of the components of the heat and water balances. This powerful index is expressed as R/LP (where R is net radiation at the Earth's surface, L is the latent heat of evaporation, and P is mean annual precipitation). Second, he used the net annual radiation balance of the Earth's surface, which characterizes the amount of energy available to power surface processes. The general zonal disposition of world's principal soil and vegetation types is defined well by isolines of R/LP: index values of 0.33 and less correspond to tundra; values from 0.33 to 1.0 correspond to forest; values from 1.0 to 2.0 correspond to steppe; values from 2.0 to 3.0 correspond to semi-desert; and values more than 3.0 correspond to desert. An even better picture of climatic influences on vegetation emerges when net radiation, R, is included as a separate factor. The principle zonobiomes – tundra, forest, steppe, semi-desert, and desert – are bounded by certain limits of the radiative index of dryness. Within the forest and steppe zonobiomes, differences in the radiation balance, R, produce marked changes of vegetation, though the general character of the vegetation remains the same.

Relations between life and climate are the subject of bioclimatology. Several schemes for characterizing bioclimates are available. Perhaps the most widely used is the 'climate diagram' devised by Walter that is explained fully in the *Klimadiagramm-Weltatlas* (Walter and Lieth 1960–67). This is the system of summarizing ecophysiological conditions that makes David Bellamy 'feel like a plant' (Bellamy 1976: 141)! Henri Gaussen and François Bagnouls's system is less well known, but has many good points (e.g. Gaussen and Bagnouls 1952; Bagnouls and Gaussen 1953, 1957; Gaussen

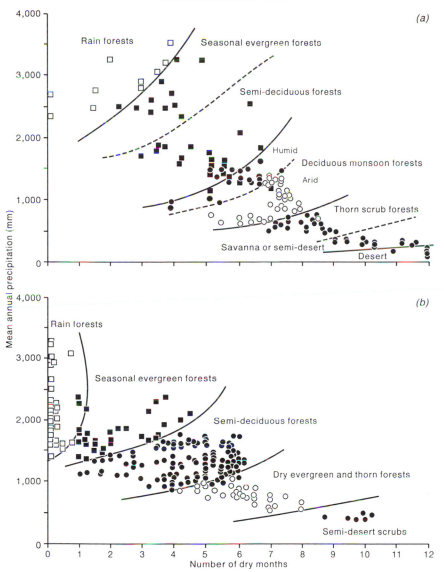

Figure 4.6 Vegetation zones within bioclimatic 'space'. (a) India. (b) Cuba
Sources: (a) After Walter (1962); (b) After Borhidi (1991)

1963). Valuable though these bioclimatic systems be, they are defective in certain particulars. They are, for example, good at predicting zonal vegetation types in very dry climates and wet climates, but are less good in seasonal climates. To overcome this problem, Walter suggested using a co-ordinate system in which the ordinate is annual precipitation and the abscissa is duration of the arid period (number of dry months). Applying this method to India, several zones of vegetation were delineated (Figure 4.6a). Using the same method on Cuban vegetation, five zonal vegetation types were disclosed (Figure 4.6b). These diagrams suggest that there are thresholds in annual precipitation and the annual distribution of precipitation that have bioclimatic significance. In Cuba, there is an upper threshold at about 1,700 mm annual precipitation above which the annual distribution has an overriding influence on the vegetation type, and a lower threshold of about 500 mm below which annual precipitation has a predominating influence on vegetation type. Between the upper and lower limits, vegetation type is a function of both annual totals and seasonal distribution. In India, the thresholds are 200 mm and 500 mm.

Another bioclimatic system, devised by Vispy-Minocher Meher-Homji (1963) using the Bagnouls and Gaussen system as a base, uses two co-ordinates representing temperature and 'moisture' values. Temperature is divided into nine classes according to mean temperatures of the warmest and the coldest months and, in the case of classes 8 and 9, the mean annual temperature (t-index). 'Moisture' is defined by a complex aridity–humidity–frost index that takes into account annual precipitation modified according

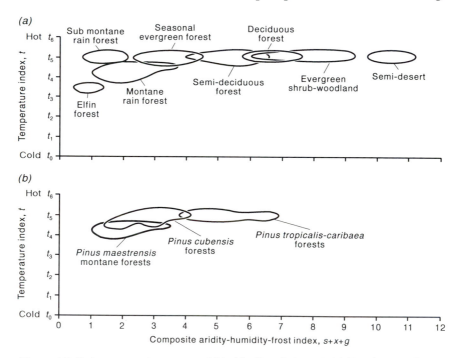

Figure 4.7 Cuban vegetation zones within bioclimatic 'space'. (a) Zonal vegetation.
(b) Pine forests and woodlands
Source: After Borhidi (1991)

to the number of dry months and periods of freezing temperatures (s-index), the number of dry months (x-index), and the number of months with frosts when physiological drought occurs (g-index). Each of the indices is assigned a value between 1 and 6, with halves included to give twelve possible values. 'Moisture' is then measured by summing the s, x, and g indices to give a composite $s + x + g$ index. Application of the Meher-Homji bioclimatic indices to Cuban vegetation leads to eight zonal vegetation belts neatly packaged within the ecological co-ordinate system (Figure 4.7a). Furthermore, by plotting the distribution of Cuban pine forests and pine woodlands within the co-ordinate system (Figure 4.7b), it was shown that different species of pine are associated with a zonal vegetation belt and are edaphic climax communities on latosols: *Pinus tropicalis* and *Pinus caribaea* belong to the semi-deciduous forest belt, *Pinus maestrensis* is associated with montane rain forests, and *Pinus cubensis* may be thought of as a paraclimax community of the seasonal-evergreen forest belt (Borhidi 1991).

Mechanisms of ecophysiological adaptation

There is a strong relationship between climatic types and vegetation types. Interesting questions focus on the mechanisms upon which these relationships are built. To investigate the correlation between vegetation types and the water balance, Nathan L. Stephenson (1990) correlated North American plant formations with actual evapotranspiration and annual water deficit (Figure 4.8). His findings are particularly informative. Tundra and forest formations occur where annual water deficits are low, roughly less than 400 mm. Formations within this band of soil–water deficit are differentiated by actual evapotranspiration to give the sequence tundra, coniferous forest, deciduous forest, and southern mixed coniferous and deciduous forest. Grasslands and shrublands are associated with high annual water deficits, mostly more than 300 mm. Annual actual evapotranspiration distinguished grassland (mostly above 300 mm) from shrubland (70–380 mm). Within the shrub formation, the transition from cold desert (Great Basin sagebrush, *Artemisia tridentata*) to warm desert (creosote bush, *Larrea tridentata*) paralleled an increase in annual water deficit. Sagebrush occurred where the deficit was less than 900 mm, the creosote bush where it exceeded 2,000 mm. Chaparral occurs at higher water deficits (800–1,070 mm) and slightly higher actual evapotranspiration (250–380 mm) than the Great Basin sagebrush (mostly 200–250 mm). Lines joining equal values of annual actual evapotranspiration and annual water deficit on Figure 4.8 would represent transects of increasing aridity at nearly constant annual energy supply. A line running from an actual evapotranspiration of 1,000 mm to an annual water deficit of 1,000 mm would represent an east-to-west transect along latitude 40° N in North America. The vegetational sequence following the humidity–aridity gradient (made famous by writers of textbooks) runs thus: deciduous

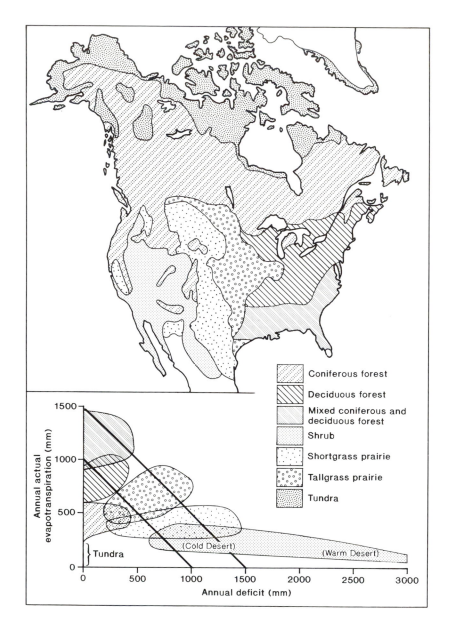

Figure 4.8 The location of major North American plant formations (top) and their relationship to annual actual evapotranspiration and the annual water deficit (bottom). For clarity, three transition formations (northern mixed forest, woodland and savanna, and shrub steppe) are not shown, but the climatic data for these accord with their bridge positions between adjacent formations

Source: After Stephenson (1990)

forest, tallgrass prairie, shortgrass prairie, and cold-desert shrub. A line joining 1,500 mm co-ordinates would represent an east-to-west transect at a lower latitude, around 30° to 35° N. By contrast, in eastern North America the annual water deficit is low and actual evapotranspiration is limited mainly by energy. Here, the dominant change of water balance is along a south-to-north transect and is dictated by a continuous fall in actual evapotranspiration. The resulting vegetational sequence is southern mixed forest, deciduous forest, coniferous forest, and tundra. A south-to-north transect in the western half of the continent, where actual evapotranspiration is low and limited chiefly by water supply, starts with a high annual water deficit that grades into a low deficit. Along this transect the vegetation changes from shrub desert, to shortgrass prairie, to coniferous forest, to tundra.

Similar work by Ronald P. Neilson and his colleagues (1992) suggests that the distribution of biomes in the United States is dependent upon the regional water balance and, in some cases, cold winter temperatures. By looking at spatial patterns of seasonal precipitation and runoff, inferences were drawn about evapotranspiration and hence vegetational relationships with climate. Taking a northern and a southern transect across the United States, they found that the eastern forested region has high precipitation in all months (Figure 4.9). As winter precipitation sharply declines, so forest gives way to grassland. The Great Plains grasslands are associated with the near absence of winter rains and the presence of spring rains. Within the grasslands, the pattern of summer and autumn rains appears to produce regional variations – as in the shortgrass prairie in Colorado or mesquite-grassland in Texas. The Great Basin sagebrush-steppe vegetation is associated with a rise in winter precipitation. The south-western deserts correspond to a near absence of spring and fall rains, relatively dry winters, and the presence or absence of summer rains. On the west coast, vegetation is typified by high winter rainfall and low summer rainfall. A steep gradient of winter rainfall (and temperature) separates the north-west coast from the south-west coast. Interestingly, the seasonal pattern of surface runoff demarcates biome boundaries even more closely than the climatic patterns. For instance, the prairie–forest ecotone corresponds to a change from year-round runoff (in the forested regions) to spring runoff (in the prairie). The only biomes in the United States showing significant summer runoff are those that receive sufficient winter precipitation to recharge deep soil-water storage, the amount of recharge appearing to compensate withdrawal by deep-rooted plants during the growing season. Nowhere does the summer rainfall balance the demand for evapotranspiration. The size of the deep soil reservoir appears to determine the specific leaf area and biomass that can be supported by evapotranspiration, and sets the boundaries between forest and grassland and grassland and shrubland. The grassland is sensitive to moisture near the soil surface, and seems directly linked to spring soil-moisture and spring rains, with summer rains allowing greater biomass and, perhaps, more

ANIMAL COMMUNITIES

Faunal zones

Homeotherms are influenced by environmental temperature only to a small degree. Unlike their poikilothermic cousins, they are not confined to warm climatic zones. Even poikilotherms are not entirely at the mercy of the elements, since they can seek refuge from severe conditions. For these reasons, the distribution of animals tends to be less influenced by climate than by historical factors, such as geological changes and evolutionary processes. In the East European and northern Asiatic region, where zones of climate and vegetation are pronounced, only a few animal species are confined to a particular climatic zone (Walter and Breckle 1985: 30–32). Those species that are confined to single climatic zones or subzones included the dwarf hamster (*Phodopus sungorus*), which lives in the steppe zone, a jerboa (*Allactaga saltator*), which lives in the semiarid zone, and a vole (*Microtus arvalis*), which is confined to the temperate zone. But the distribution of many species crosses zonary borders. The reindeer (*Rangifer tarandus*) ranges from the Arctic zone to the taiga zone, the roe deer (*Capreolus capreolus*) from the temperate zone to the steppe zone. Few studies of this kind have been engaged upon. Such evidence as exists suggests that the distribution of the fauna, unlike that of the flora, seldom correlates closely with climatic zones and zonobiomes (Walter and Breckle 1985: 32). That being so, it is reasonable to conclude that world faunation is not arranged on climatic lines, but consists of a series of regions, roughly corresponding to the modern continents, that have resulted primarily from geological changes (e.g. C. H. Smith 1983a, 1983b; Mayr 1965). This contrasts with world vegetation which, though it has been subject to the same geological influences, has proved far more responsive to the influence of climate.

Work on the avifauna of Northern Territory, Australia, tends to disaffirm the view that animals are not zoned on climatic lines (Whitehead *et al.* 1992). Presence and absence data on bird species in 129 grid-cells 1° × 1° (latitude × longitude) in size, covering the whole of mainland Northern Territory and its major islands, was subjected to TWINSPAN analysis. Eight geographically coherent regions were uncovered up to the third division of TWINSPAN, all of which were arranged in latitudinal bands. The four avian zones defined by primary and secondary divisions of TWINSPAN are: the central arid zone, the dry grassland zone, the northern savanna, and the wet tropical zone (Figure 4.10; Plate 4.1). There is a broad conformity between the avifaunal zones established using multivariate methods and other studies that used numerical methods to locate regional boundaries, but not with the biogeographical zones delimited by Miklos D. F. Udvardy (1982) (Figure 4.11). Caution is necessary when interpreting the results of classification techniques because it is possible that they will force inherently continuous

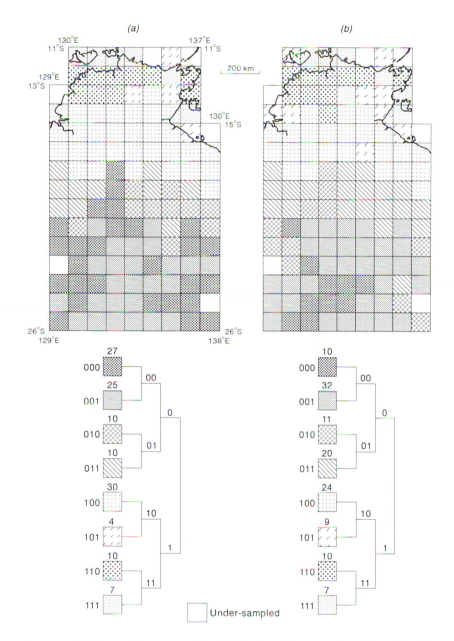

Figure 4.10 Avian species composition in Northern Territory, Australia, within 1° × 1° (latitude × longitude) grid-cells, classified by TWINSPAN. (a) Land birds. (b) All birds. Grid-cells with less than twenty species reported, and species occurring in less than five grid-cells, were excluded
Source: After Whitehead *et al.* (1992)

Plate 4.1a Typical vegetation in the wet tropical avian zone of the Northern Territory, Australia, as recognized by Whitehead *et al.* (1992): *Eucalyptus tetrodonta* tall tropical grass savanna. This is the dominant dryland plant community on the north coast. Note, however, that monsoon rain forest occurs in small scattered isolates in the vast expanse of the *Eucalyptus* savanna and large areas of freshwater floodplains occur on the major rivers.
Photograph by David Bowman

Plate 4.1b Typical vegetation in the northern savanna avian zone of the Northern Territory, Australia, as recognized by Whitehead *et al.* (1992): savannas in the Victoria River District. These savannas are dominated by *Eucalyptus*, *Lisiphyllum*, and *Terminalia*.
Photograph by David Bowman

Plate 4.1c Typical vegetation in the dry grassland avian zone of the Northern Territory, Australia, as recognized by Whitehead *et al.* (1992): treeless grassland of the Barkly Tablelands. The dominant grass species is *Astrebla pectinata*. Deep cracking clays are thought to prevent the growth of trees.
Photograph by David Liddle

Plate 4.1d Typical vegetation in the central arid avian zone of the Northern Territory, Australia, as recognized by Whitehead *et al.* (1992): *Triodia clelandii* hummock grasslands with mixed species of scattered shrubs including *Eremophila*, *Hakea*, *Acacia*, and *Eucalyptus* on the MacDonnell Ranges in central Australia.
Photograph by David Bowman

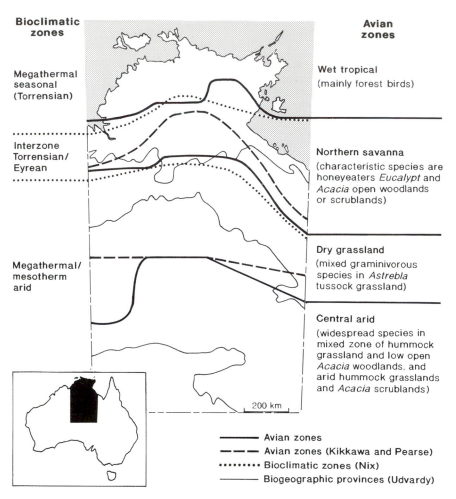

Bioclimatic zones

Megathermal seasonal (Torrensian)

Interzone Torrensian/ Eyrean

Megathermal/ mesotherm arid

Avian zones

Wet tropical
(mainly forest birds)

Northern savanna
(characteristic species are honeyeaters *Eucalypt* and *Acacia* open woodlands or scrublands)

Dry grassland
(mixed graminivorous species in *Astrebla* tussock grassland)

Central arid
(widespread species in mixed zone of hummock grassland and low open *Acacia* woodlands, and arid hummock grasslands and *Acacia* scrublands)

200 km

——— Avian zones
– – – Avian zones (Kikkawa and Pearse)
········ Bioclimatic zones (Nix)
——— Biogeographic provinces (Udvardy)

Figure 4.11 A comparison of primary and secondary avian zones (bold solid lines) and three other zonal groupings as they affect the Northern Territory: Kikkawa and Pearse's (1969) avifaunal regionalization (bold dashed line); Nix's (1982) bioclimatic regionalization (bold dotted line); and Udvardy's (1982) scheme of biogeographical provinces (thin line)
Source: After Whitehead *et al.* (1992)

variation into discrete groups. The primary division picked out by TWIN-SPAN lies between the 17th and 18th parallels, separates dry grassland from northern savanna, and delimits a true discontinuity in avian species composition. All other shifts in avian species composition are incremental, or very nearly so. Detrended correspondence analysis (DCA) was used to retrieve the broad environmental correlates of the data. The avifaunal zones mirror a steep climatic gradient running from the northern seasonally wet tropics to arid central Australia. Scores on the first axis generated by the DCA are strongly correlated with grid-cell latitude ($r = -0.94$) and with grid-cell elevation ($r = -0.75$). Ordination scores of the first three axes all displayed significant correlation with climatic variables, the strongest correlation being between the first axis and rainfall in the wettest quarter ($r = -0.95$). Rainfall variables tended to show greater within-group homogeneity and larger among-group variance than did temperature variables. Rainfall in the wettest quarter showed the largest among-group variance, and rainfall in the coolest quarter the least. The DCA suggests environmental relationships but does not explain them. Climate may exercise an influence on avian species distribution directly through physiological constraints, or indirectly through the availability and quality of habitat, or through a mix of direct and indirect effects. But it is indisputably the case that, in the study area, vegetation patterns, and thus the distribution of habitats available to birds, are strongly influenced by climate.

Clines

Animal species are adapted to conditions in their local environment. Many animal species adapt to gradual geographical changes in climate. Such adaptation is often expressed in the phenotype as a measurable change in size, colour, or some other trait. The gradation of form along a climatic gradient is called a cline (Huxley 1942). Clines result from local populations developing tolerances to local conditions, including climate, through the process of natural selection. Such local populations constitute an ecological race. Ecological races may display either gradual or abrupt change along an environmental gradient, and these racial gradients are called ecoclines. Clinal variation in animal populations produced by regular gradients of climatic factors, notably temperature and moisture, might be designated climoclines. This would serve to distinguish them from clines created by other environmental factors, which might be termed lithoclines, pedoclines, or some other kind of cline, as appropriate.

Morphological climoclines may evolve very swiftly. In 1852, the house sparrow (*Passer domesticus*) was introduced into the eastern United States from England. Fifty years later it had already developed geographical variation in size and colour. Today it is smallest along the central Californian coast and south-east Mexico, and is largest on the Mexican Plateau, the Rocky Mountains, and the northern Great Plains. The clines in house sparrows that have evolved in North America resemble the clines found in Europe (Johnston and Selander 1971). The American robin (*Turdus migratorius*) displays similar geographical variation in size and shape to the house sparrow (Aldrich and James 1991). It is small in the south-eastern United States and along the central Californian coast, and large in the Rocky

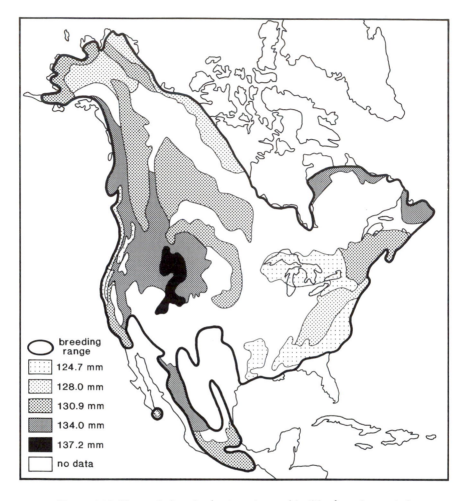

Figure 4.12 Size variations in the American robin (*Turdus migratorius*).
The units are median wing length of males measured in thirty-six ecological regions
during the breeding season. The bold line outlines the breeding range of the robin.
Blank areas lack data
Source: After Aldrich and James (1991)

Mountains and associated high plains (Figure 4.12). Clinal variation in the European wild rabbit (*Oryctolagus cuniculus*), introduced in eastern Australia a little over a century ago, already displays clinal variation in skeletal morphology. The rapidity of clinal evolution revealed by these empirical studies has been reproduced using genetic models of populations which show that, even in the presence of gene flow, clines can develop within a few generations (Endler 1977).

To explain the presence of climoclines, it is common to resort to

90

ecogeographical rules. Most of these were established during the nineteenth century when it was observed that the form of many warm-blooded animal species varies in a regular way with climate. The first ecogeographical rule was established by Constantin Wilhelm Lambert Gloger in 1833. Gloger's Rule states that races of birds and mammals in warmer regions are more darkly coloured than races in colder or drier regions. Or, to put it another way, birds and mammals tend to have darker feathers and fur in areas of higher humidity. This is recognized as a valid generalization about clines of melanism. An explanation for it is that animals in warmer, more humid regions require more pigmentation to protect them from the light. Gloger's rule was first observed in birds, but later seen to apply to mammals such as wolves, foxes, and hares. It has also been observed in beetles, flies, and butterflies. Given that colour variation shows a concordance of pattern in birds that have vastly different competitors, diets, histories, and levels of gene flow, some common physiological adaptation seems likely.

Bergmann's Rule, also known as the Size Rule, states that species of birds and mammals living in cold climates are larger than their congeners that inhabit warm climates. It applies to a wide range of birds and mammals, and was established by Carl Bergmann in 1847. Bergmann believed many species conform to the rule because being bigger confers a thermal advantage on individuals in cold climates: as an object increases in size, its surface area becomes relatively smaller (increasing by the square) than its volume (increasing by the cube). Examples of Bergmann's Rule, and exceptions to it, abound. In central Europe, the larger mammals, including the red deer, roe deer, bear, fox, wolf, and wild boar, increase in size towards the north-east and decrease in size towards the south-west. The skull length in the wild boar ranges from 560 mm in Siberia to 324 mm in southern Spain. Species that decline to obey Bergmann's rule include the capercaillie (*Tetrao urogallus*), which is smaller in Siberia than in Germany. Also, many widespread Eurasian and North American bird species are largest in the highlands of the semiarid tropics (Iran, the Atlas Mountains, and the Mexican Highlands), and not in the coldest part of their range. The geographical variation of size in some vertebrate and invertebrate poikilotherm species conforms with Bergmann's rule (Lindsey 1966; Ray 1960; Rensch 1932).

Ernst Mayr (1956) revised Bergmann's Rule to apply only to races within species, and dropped the argument about thermal adaptation, the physiological grounds for which are rather shaky: it is doubtful whether size change with ambient temperature is a consequence of heat conservation. The relative reduction in surface area is too small for an effective reduction in heat loss. Insulation and vascular control are far more efficient ways of regulating body temperature. As a large animal will have a greater overall energy requirement than a small animal, large size may reflect the ability to store energy, a factor that would seem to be rather crucial in a harsh environment. Other factors have been educed to explain Bergmann's Rule. These include the presence or absence of potential competitors, food size, habitat productivity, species diversity, and equilibrial niche size (e.g. Rosenzweig 1966, 1968; Boyce 1978; Koch 1986). None the less, recent studies, such as those discussed below, strongly indicate that climate plays a leading role.

A new version of Bergmann's Rule was proposed by Frances C. James (1970). She found a negative relationship between size in birds and various climatic measures sensitive to both temperature and moisture (wet-bulb temperature, vapour pressure, and absolute humidity) in the eastern and central United States. Wing length, a good surrogate of body size, increased in size northwards and westwards from Florida in the following species: the hairy woodpecker (*Dendrocopos villosus*), downy woodpecker (*Dendrocopos pubescens*), blue jay (*Cyanocitta cristata*), Carolina chickadee (*Parus carolinensis*), white-breasted nuthatch (*Sitta carolinensis*), and eastern meadowlark (*Sturnella magna*). In all cases, there was a tendency for larger (or longer-winged) birds to extend southwards in the Appalachian Mountains, and for smaller (or shorter-winged) birds to extend northwards in the Mississippi River valley. In the downy woodpecker, female white-breasted nuthatches, and female blue jays, relatively longer-winged birds tended to extend southwards into the interior highlands of Arkansas, and relatively shorter-winged birds to extend northwards into other river valleys. These subtle relations between intraspecific size variation and topographic features indicated that the link between the two phenomena may involve precise adaptations to very minor climatic gradients. The variation in wing length in these bird species correlated most highly with those variables, such as wet-bulb temperature, which register the combined effects of temperature and humidity. This suggested that size variation depends on moisture levels as well as temperature. James reasoned that a relationship with wet-bulb temperature and with absolute humidity in ecologically different species strongly suggests that a common physiological adaptation is involved. Absolute humidity nearly determines an animal's ability to lose heat: any animal with constant design will be able to unload heat more easily if it has a higher ratio of respiratory surface to body size. This new twist to Bergmann's Rule bolsters some aspects of Bergmann's original interpretation about thermal budgets. Climate tends to be cooler, and therefore drier, at high altitudes and latitudes. This accounts for the fact that many clines of increasing size parallel increasing altitude and latitude. Additionally, size tends to increase in arid regions irrespective of altitude and latitude, and widespread species tend to be largest in areas that are high, cool, and dry. James concluded that, if the remarkably consistent pattern of intraspecific size variation in breeding populations of North American birds represents an adaptive response, then 'Bergmann's original rationale of thermal economy, reinterpreted in terms of temperature and moisture rather than temperature alone, still stands as a parsimonious explanation' (James 1991: 698).

Recent work on geographical variation within mammal populations has

Plate 4.2 The coyote (*Canis latrans*).
Photograph by Michael L. Kennedy

revealed that size is influenced by moisture, as well as by temperature. The coyote (*Canis latrans*) is widely distributed in North America (Plate 4.2). A study of individuals living in the south-eastern United States showed that in both males and females, larger individuals occur in the more eastern localities and smaller animals occupy the more western localities (Kennedy *et al.* 1986). This was revealed by principal component analysis of fourteen male and twelve female skull measurements. The first three components accounted for 87.9 per cent of the phenetic variation in male skulls and 94.1 per cent of the phenetic variation in female skulls. Three-dimensional projections of localities on to the principal components showed that, for both sexes, larger individuals occurred in more easterly locations (Figure 4.13). Large size was positively correlated with high actual evapotranspiration, except in localities with the highest evapotranspiration. This accorded with Michael L. Rosenzwieg's (1968) finding that actual evaporation is a good predictor of size, particularly in areas of low productivity, but if actual evapotranspiration is very high, then it will either be uncorrelated or negatively correlated with body size. Not all investigations of size variation reveal such clear results. Morphometric variation in fourteen species of kangaroo rat (genus *Dipodomys*) revealed little correlation with latitude, annual range of temperature, seasonal range of temperature, precipitation, or actual evapotranspiration (Baumgardner and Kennedy 1993). A tendency for size to increase with decreasing annual temperature was noted. It is possible that an examination of fewer species of kangaroo rats might have led to different conclusions. The

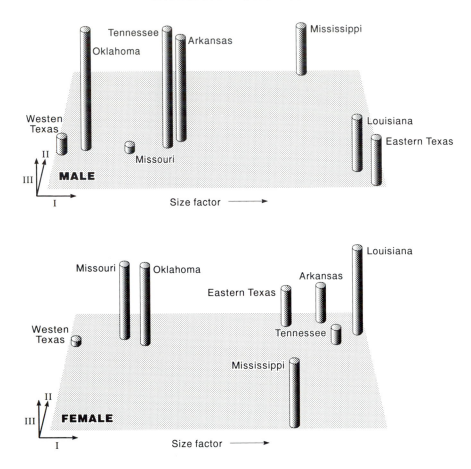

Figure 4.13 Eight male and eight female localities of the coyote (*Canis latrans*) in the southern United States projected into the space created by the first three principal components extracted from a correlation matrix of skull characters (fourteen male, twelve female). Axis I is a size factor in both sexes and accounts for 65 per cent (male) and 73.4 per cent (female) of the variation in the data. Axis II has high loadings with width of upper molar two and orbital length in males (localities in which males have relatively wide upper molar twos and long orbital lengths are placed near the front of the diagram, and vice versa), and has high loadings with skull length at condyles and canine diameter in females (localities in which females have relatively short skull heights at condyles and short canine diameters are placed at the front of the diagram, and vice versa). Axis III has highest loadings with braincase width in males, and with width between alveoli of the palatines in females (localities in which males have a wide braincase, and females have a wide distance between alveoli of the palatines, are placed on longer 'rods')
Source: After Kennedy *et al.* (1986)

indications are that abiotic factors may influence morphometric variation in some parts of a species' range, but may be supplanted by other factors elsewhere.

There are other ecogeographical rules. Allen's Rule, or the Proportional Rule, extends Bergmann's Rule to include protruding parts of the body, such as necks, legs, tails, ears, and bills. Joel A. Allen (1877), after whom the rule is named, found that protruding parts in wolves, foxes, hares, and wild cats are shorter in cooler regions. Like large body size, short protruding parts help to reduce the surface area and so conserve heat in a cold climate. The jackrabbit (subgenus *Macrotolagus*), which lives in the south-western United States, has ears one third its body length; in the common jackrabbit (*Lagus campestris*), which ranges from Kansas to Canada, the ears are the same length as the head. Another observation conforming to Allen's Rule is that such mammals as bats, which have a large surface area for their body mass, are found chiefly in the tropics. Allen's Rule has been observed in poikilotherms as well as homeotherms. Hesse's Rule, also known as the Heart-Weight Rule, was proposed by Richard Hesse (e.g. Hesse *et al*. 1937). Like Allen's Rule, it is an extension of Bergmann's Rule. It states that extra metabolic work done to maintain heat in a cold environment causes a greater volume and mass of heart in animals living there as compared with their counterparts in warmer regions. Thus, the relative heart weight of the sparrow (*Passer montanus*) is 15.74 per thousand in St Petersburg. 14.00 per thousand in northern Germany, and 13.1 per thousand in southern Germany. Ernst Mayr (1942) added a few extra rules that apply only to birds: in colder climates, the number of eggs in a clutch is larger, the digestive and absorptive parts of the gut are larger, the wings longer, and migratory behaviour is more developed.

Ecogeographical regularities in the size, colour, and other characteristics of a species have until recently been looked at individually. In reality, all characteristics of a species will respond to climate in concert: climate wields a strong influence on a suite of life-history characteristics. In North America, the litter size of small mammals tends to increase with latitude while the number of litters per year decreases (e.g. Lord 1960; Cameron and McClure 1988), and body size tends to increase northwards, so conforming to Bergmann's Rule (Cameron and McClure 1988). This is clearly shown in the relationship between litter size and latitude in muskrat (*Ondatra zibethicus*) populations in North America (Figure 4.14). Notice that the relationship is curvilinear with an asymptote at about 56.6°N. Climatic influences on the life history characteristics of North American muskrats are fairly well established (Boyce 1978). A recent study of a northern and southern population shows the multivariate nature of clinal variations (Simpson and Boutin 1993). The northern population lives at Old Crow Flats, Yukon Territory (68° 05′N), the southern population in Tiny Marsh, Ontario (44° 35′N). They were both studied during the summers of 1985 and 1986.

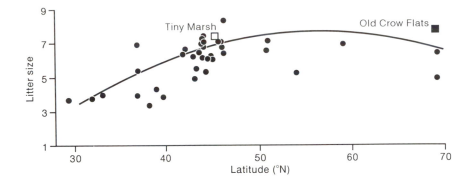

Figure 4.14 Relationship between litter size and latitude in the North American
muskrat (*Ondatra zibethicus*)
Source: After Simpson and Boutin (1993)

Life history traits of the two populations show significant similarities and
differences. Litter size was roughly the same, but average annual productivity
was greater at Tiny Marsh where individuals produced more litters per
season. The similarity of litter size, despite the large difference in latitude, is
predicted from the relationship expressed in Figure 4.14. Both the growth
rates and the weight of juvenile and adult muskrats were appreciably greater
at Tiny Marsh, probably because highly nutritious food is available in the
Ontario marshes over a long growing season. However, if the abundance and
quality of food resources lead to difference in reproduction and body weight,
then survival over winter might be expected to be lower in the northern
population. As it happened, the number of juvenile muskrats surviving the
winter was greater at Old Crow Flats than at Tiny Marsh. This may be
because the shorter growing season and local energetic limitations on growth
and reproduction promote a conservative reproductive strategy in the
northern population, geared to enhance survival over long periods of low
food availability; whereas, a long growing season and high-quality food
resources allows a more productive reproductive strategy in the southern
population. This adaptive response seems to affect body size: members of the
northern population tend to be smaller than members of the southern
population. Thus, in this case, low food availability appears to override the
direct influence of climate and Bergmann's Rule is not upheld.

An outstanding example of a species' overall adaptive response to climate
is provided by subterranean mole rats of the *Spalax ehrenbergi* complex
living in Israel (Nevo 1986). These mole rat populations comprise four
morphologically identical incipient chromosomal species (with diploid
chromosome numbers 2n = 52, 54, 58, and 60). The four chromosomal

species appear to be evolving and undergoing ecological separation in different climatic regions: the cool and humid Galilee Mountains (2n = 52), the cool and drier Golan Heights (2n = 54), the warm and humid central Mediterranean part of Israel (2n = 58), and the warm and dry area of Samaria, Judea, and the northern Negev (2n = 60). All the species are adapted to a subterranean ecotype: they are little cylinders with short limbs and no external tail, ears, or eyes. Their size varies according to heat load, presumably so that there is only a small risk of overheating under different climates: large individuals live in the Golan Heights; smaller ones in the northern Negev. The colour of the mole rats' pelage ranges from dark on the heavier black and red soils in the north, to light on the lighter soils in the south. The smaller body size and paler pelage colour associated mainly with 2n = 60 helps to mitigate against the heavy heat load in the hot steppe regions approaching the Negev desert. The mole rats show several adaptations at the physiological level. Basal metabolic rates decrease progressively towards the desert. This minimizes water expenditure and the chances of overheating. More generally, the combined physiological variation in basal metabolic rates, non-shivering heat generation, body-temperature regulation, and heart and respiratory rates, appears to be adaptive at both the mesoclimatic and microclimatic levels, and both between and within species, so contributing to the optimal use of energy. Ecologically, territory size correlates negatively, and population numbers correlate positively, with productivity and resource availability. Behaviourally, activity patterns and habitat selection appear to optimize energy balance, and differential swimming ability appears to overcome winter flooding, all paralleling the climatic origins of the different species. In summary, the incipient species are reproductively isolated to varying degrees representing different adaptive systems that can be viewed genetically, physiologically, ecologically, and behaviourally. All are adapted to climate, defined by humidity and temperature regimes, and ecological speciation is correlated with the southwards increase in aridity stress. Whether climate plays such a dominant role in the evolution of most animal species has yet to be discovered. It seems likely.

SPECIES RICHNESS GRADIENTS

Studies on several groups of animals and plants have established latitudinal, altitudinal, and (to a lesser extent) longitudinal gradients of species richness. Such studies have also revealed a tendency for fewer animal species to occur on peninsulas. A primary aim of modern ecology has been to discover the factors that are responsible for producing this pattern. Historical processes (speciation and dispersal), climate, climatic variability, topography, biotic processes (primary productivity, competition, and so forth), disturbance, and the richness of other groups of organisms have all been suggested as important factors in explaining species richness. There is little doubt that all these factors can operate on a local scale. A number of recent studies have shown clearly that, for both animals and plants, present-day species richness for largish regions can be explained very well using climatic factors, especially available energy.

A species–energy hypothesis of species richness was expounded by David Hamilton Wright (1983). Wright's hypothesis states that, where water supply and other environmental factors are not limiting, diversity within terrestrial habitats is largely controlled by the amount of solar energy available, declining latitudinally in line with the polewards decrease of solar radiation receipt. In support of his hypothesis, Wright produced highly significant regression equations in which plant and bird diversity from a world-wide sample of thirty-six islands is related to solar energy. Against Wright's hypothesis, it might be objected that solar energy and diversity are both known to decline with latitude and one would expect them to be correlated. However, a study of the diversity of butterflies and moths in Great Britain showed that the species–energy relationship holds only during that part of the year when insects are absorbing energy, and cannot be detected during diapause; and that diversity is highly correlated with sunshine and temperature data, variables which are semi-independent of latitude (J. R. G. Turner et al. 1987). A later study of small insectivorous birds in Britain avoided the problem of picking up the latitudinal decrease in species richness by hypothesizing that a correlation between energy and species diversity would only apply during winter months when organisms are actively absorbing energy. It found that the varying bird distribution in summer and winter corresponds well with the prediction of the species–energy theory (J. R. G. Turner et al. 1988). This work has met with stiff opposition from some quarters (see Cousins 1989; Elkins 1989; J. R. G. Turner and Lennon 1989).

As a test of the species–energy theory, the spatial pattern of tree species richness in North America was investigated by David J. Currie and Viviane Paquin (1987). Only in the east of North America does the richness of tree species display a marked latitudinal gradient (Figure 4.15). Maximum tree species richness occurs on the high plateau of the Appalachian Mountains, whereas minimum tree species richness occurs in areas immediately to the east of the Rocky and Sierra Nevada Mountains. Peninsulas do not contain notably fewer species. This pattern contrasts with species richness patterns for birds and mammals where maxima occur in the Rockies and Appalachians, local minima occur on peninsulas, and fewer species occur in the south-east (Figure 4.15). In an attempt to account for the richness pattern of trees, species richness was correlated with physical and environmental factors. The strongest correlation was with annual evapotranspiration, this variable accounting for 76 per cent of the variation in species richness. Most of the residual variation, explored by multiple regression, was explained by range of elevation within a quadrat and by distance from the coast, especially in low latitudes. No significant variation in the data could be attributed to

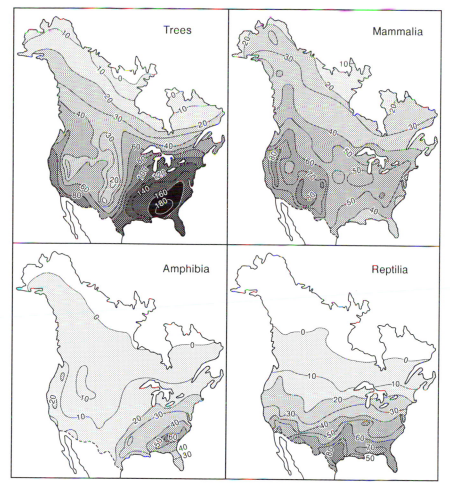

Figure 4.15 The species richness of North American trees, mammals, amphibians, and reptiles. Contours join points with the same approximate number of species per quadrat
Source: After D. J. Currie and Paquin (1987) and D. J. Currie (1991)

quadrat area, seasonality of climate, whether the quadrat had been glaciated, or whether the quadrat were on a peninsula. A crucial question is why evaporation should be so important in explaining tree species richness. A likely answer is that evapotranspiration is closely connected with terrestrial primary productivity and is thus a measure of energy consumption by a community. This idea is consistent with the hypothesis that energy is partitioned among species and the total available energy thus limits the number of species in an area. Species richness is influenced by other factors.

Mountains, for example, being physically complex, allow even more partitioning of energy than the available energy level would suggest and so can support a greater number of species.

A surprising result of the work on British birds and North American trees is the apparent insignificance of historical processes such as glaciation and dispersal. To explore this finding more fully, an investigation was made of tree species richness in Great Britain and Ireland. The accepted view is that the British and European flora are impoverished in comparison with the North America flora owing to post-glacial barriers restricting recolonization in Europe. However, the observed tree species richness in the British Isles is very close to the number predicted by the regression equations established by analysis of the North American data. It would be rather unwise to discard historical factors out of hand. Nevertheless, it would seem that the most parsimonious interpretation of the results is that tree species richness in North America and the British Isles varies among regions primarily as a function of varying amounts of energy available to the community (D. J. Currie and Paquin 1987).

In a further study, Currie (1991) mapped the number of species of mammals, reptiles, amphibians, and birds in North America (Figure 4.15; bird species richness not shown) and correlated them with twenty-one descriptors of the environment. Environmental variables were chosen to test a variety of hypotheses concerning species richness. The maps reveal the well-known latitudinal decline in species richness. In the cases of reptilia and trees, the richness–latitude relationship is monotonic. But in the cases of birds and mammals, pronounced richness peaks occur at latitudes 44°N and 39°N respectively, while amphibian richness peaked more gently at latitude 34°N. The pattern of tree richness has already been discussed. It would seem reasonable to suppose that the richness of animals would vary as a function of the richness of vascular plants, which itself should covary with tree species richness. However, it is evident from Figure 4.15 that only amphibian richness shows a clear monotonic relation with tree species richness. It is possible that animal species richness is related to primary productivity, but, except in the case of the amphibians, no tight relationship between species richness and productivity was apparent in the study. By far the strongest relations, as measured by non-parametric correlation coefficients, were between vertebrate species richness and virtually all the climatic variables, with the three strongest correlates being annual potential evapotranspiration, solar radiation, and mean annual temperature, all of which are aspects of the regional energy balance. Annual potential evapotranspiration, which is a measure of crude, integrated, ambient energy, alone accounted for about 79 per cent of variability in species richness. Even at the level of family and order, richness was highly correlated with annual potential evapotranspiration. Thus, the outcome of this study is a strong vindication of the species–energy hypothesis: of all the factors that could affect species richness, energy (or, at least, surrogate climatic measures of it) stands out as the chief factor. This is fairly easy to appreciate in the case of plants. According to the species–energy hypothesis, in regions of the same area, species richness is determined by the energy flux. Primary productivity in plants is a direct measure of energy capture and, on a large scale, correlates most strongly with annual potential evapotranspiration. The case of animal species richness is less straightforward because several hypotheses should be considered. However, Currie's researches suggest that the only tenable hypothesis at the scale considered is that animal species richness is determined largely by regional ambient energy levels.

But why should vertebrate richness be influenced more by crude atmospheric energy than by energy available in the form of food? A possible answer is that the thermal budget of individuals, which depends on atmospheric energy, is more important in determining species richness than is food energy. Another speculative answer was proffered by Currie: so long as water is not limiting, both plant species richness and animals species richness increase; but, when water is limiting and potential evapotranspiration rises above actual evapotranspiration, animal species richness continues increasing whereas plant species richness does not. So do environments in which potential evapotranspiration exceeds actual evapotranspiration offer energy sources unrelated to local primary productivity? Further work is needed to answer this question. Clearly, the energy hypothesis is not without problems, but it is a stimulating idea that gives a fresh slant on species richness patterns at a regional scale. Moreover, macroscale relationships between populations and environmental factors have been discovered (D. J. Currie and Fritz 1993). Animal population density is inversely related to potential evapotranspiration: after correcting for body weight and metabolic group (invertebrate ectotherm, vertebrate ectotherm, and various vertebrate endotherms), populations are denser in low-energy, that is high-latitude, environments. Second, population energy use increases slightly with latitude, varying less than an order of magnitude from equator to poles. As with species richness, spatial variations in population density and energy use appear to relate factors that vary geographically and that appear to be nicely summarized by environmental energy levels.

SUMMARY

Species and communities respond to the changing environmental fields in which they live. Individual species adapt to environmental factors, their geographical ranges being constrained by their limits of tolerance. Many species are undoubtedly adapted to climatic fields. Towards the edge of their geographical ranges they show signs of stress caused by climatic factors. None the less, species at the edge of their geographical ranges will be affected by all environmental factors, and not just climate. Individual species' distributions are the result of multivariate interactions and require multivariate techniques to unravel their intricacies. The same is true of broad-scale studies which synthesize information on the mesoscale distribution of several species. The value of a multivariate approach in broad-scale studies is revealed by work on *Salix* species in Europe. Plant and animals communities are subject to the same environmental influences as their component species. At a megascale, climate has a predominating influence on plant and, to a lesser extent, animal communities. Plants of different ancestry, growing under the same climatic type, possess similar physiognomic adaptations – they have a similar leaf form, for example. Plant formation-types (zonobiomes if animals are included) are communities adapted to distinct climates. Recent work focuses on the ecophysiological mechanisms by which plant communities become adapted to regional climates. Emphasis is laid upon the energy and water balances at the Earth's surface, and particularly to actual evaporation and the annual water deficit. Interestingly, palaeoecological work has shown that community response to climatic change is the outcome of adaptation of individual species: communities are constantly evolving as component species regroup in answer to environmental, and especially to climatic, change. This supports the evolutionary view of communities implicit in the 'brash' equation. Animal communities are less constrained by climatic fields than plant communities; though, as the study of avian communities in Northern Territory, Australia, suggests, some constraints do exist. Many animal species exhibit climoclinal variations, possibly explicable in terms of Bergmann's Rule, as modified by Frances C. James. Clinal variations in animal populations are influenced by many factors in the environmental complex; indeed, much recent work on microgeographical variation uses multivariate methods. Megascale and macroscale gradients of species richness, both in animals and plants, have recently been found to conform to climatic fields, and particularly to fields of available solar energy and annual evapotranspiration.

FURTHER READING

The best introduction to climate and life must be any of Walter's masterworks on the vegetation of the Earth. Woodward's *Climate and Plant Distribution* (1987) merits attention, but it is as well to see where Woodward's work has led by reading the paper by Prentice *et al.* (1992). The book by David M. Gates called *Climate Change and Its Biological Consequences* (1993) contains much valuable and up-to-date information on the relationships of plants and plant formations to climate. Many interesting ideas are also to be found in Daniel B. Botkin's book, *Forest Dynamics: An Ecological Model* (1993). No book is devoted to the species–energy theory. Interested readers must seek out the relevant papers or the discussion in Huggett (1991). Readers unfamiliar with the multivariate statistical methods mentioned in the chapter – for instance direct and indirect gradient analysis – should consult Robert H. Whittaker (1967) and more recent papers for a description of the methods.

5

ALTITUDE

It seems reasonable to suppose that climate is a primary determinant of altitudinal changes in geoecosystems, and especially of vegetation and soil zones. The general pattern of climatic changes with elevation would surely lead to this supposition (e.g. Barry 1992). On ascending a mountain, the air temperature will fall at a rate of about 6 °C every 1,000 m. Considerable deviations from this average rate will arise from local topographic and climatic conditions, and the rate is normally less in winter than in summer, and less at night than during the day. On high mountains, temperatures regularly fall below freezing point. With adequate precipitation, high mountains are therefore commonly capped with snow. On average, mountains receive more rainfall than lowlands at the same latitude. Indeed, in deserts, they may support altitudinal oases (Cloudsley-Thompson 1975b). To some extent, the differences between coastal and interior continental climates apply to the climates of coastal and interior mountain ranges. It is also generally true that the vegetation zones on large and non-isolated mountains occur higher than they do on small and isolated mountains. This is a consequence of the Massenerhebung Effect whereby great mountain masses to some extent create their own climates.

Altitudinal changes of climate should sort and sift species according to their climatic tolerance ranges, and so produce elevational belts of vegetation in the same way that latitudinal differences in climate produce latitudinal bands of vegetation. Similarly, altitudinal changes of climate should affect soil processes and disrupt the zones of 'normal' soils. The parallel between latitudinal and altitudinal temperature zones is valid, but it should be stressed that conditions in analogous zones are very different. At high altitudes there is a reduction in air pressure. This has physiological effects on the fauna. And, more significantly, in the polar zone, sunlight during the long summer day is diminished because it passes obliquely through a great thickness of atmosphere. In the alpine zone, sunlight is intense as it passes through a thin and very clear atmosphere. The wind circulation in the major Himalayan valleys involves air subsiding over valley bottoms. Consequently, valley-side slopes and ridges are relatively wet, while valley bottoms are relatively dry

Plate 5.1 Vegetation zones in a Karakoram Himalayan valley, Pakistan. Owing to the 'Troll effect', the valley bottoms are relatively dry and support only grasses, while the higher valley-side slopes are wet enough to sustain tree growth.
Photograph by David N. Collins

(the 'Troll effect'). These climatic differences, which have no direct latitudinal counterpart, are reflected in plant growth (Plate 5.1). Altitudinal and latitudinal soil sequences also differ because high altitudes are commonly areas of steep slopes whereas high latitudes contain much flat terrain. Accordingly, the parallels between latitudinal belts of soil and vegetation and their compressed altitudinal versions on mountains should not be drawn too closely: mountain landscapes are in many ways quite different from their lowland counterparts. To complicate the picture further, altitudinal effects are themselves influenced by latitude: solar radiation, net radiation, and temperature generally decline with increasing latitude. Consequently, the elevation of the tree-line and the snow-line decrease polewards, and the nival zone of permanent snow and ice occupies the tops of much lower mountains in high latitudes than in low latitudes. Additionally, seasonal and diurnal climatic cycles are more pronounced in higher latitudes. In turn, latitudinal influences on the temperature regime affect the character of precipitation in mountains. On high mountains in the equatorial zone, snow may fall on any day of the year, especially during the night. In middle and high latitudes, there is a distinct and protracted winter season.

SPECIES AND ALTITUDE

Studies of individual species' distributions with increasing elevation tend to show continuous variation, with species' distributions overlapping. David W. Shimwell (1971) studied an altitudinal transect up the mountain Corserine in the Rhinns of Kells Range, south-west Scotland (Figure 5.1). *Molinia caerulea* was the dominant species up to 460 m but did not grow above 520 m. Above 490 m, *Festuca rubra* and *Nardus stricta* were co-dominant up to 780 m. *Carex bigelowii* ranged from 600 to 820 m but did not attain dominant status. Above 780 m *Salix herbacea* rapidly became dominant.

Much work has gone into elucidating elevational influences on species' distributions. Helmut Gams (1931), working on vegetation in the Alps, found that hygric continentality was an influential determinant of forest type. Hygric continentality is the ratio of altitude, Z (m), to annual precipitation, P (mm), and may be expressed as $\tan^{-1} Z/P$, which is 45° when $Z/P = 1$. Figure 5.2 shows the relation of forest type (as defined by dominant or co-dominant tree species) in the Alps to hygric continentality (Aulitsky et al. 1982). The distribution of these forests is influenced by other environmental factors, including the duration of snow cover and mean annual temperature range (Figure 5.3). It can be seen that larch and stone-pine forest occurs on sites with high hygric continentality, but modest annual temperature range, and long-lasting snow cover. At the other extreme, beech–oak forest occurs with low hygric continentality, large annual temperature range, and brief periods of snow cover. Larch–spruce forest is intermediate between the other two.

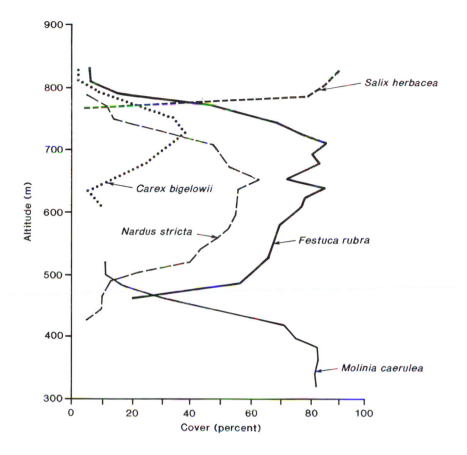

Figure 5.1 An elevational transect up the mountain Corserine, Rhinns of Kells
Range, south-west Scotland
Source: After Shimwell (1971)

A study of elevational effects on plant species in the central Rocky
Mountains was particularly penetrating (Neilson and Wullstein 1983, 1986)
(Figure 5.4). The upper elevational boundaries tend to be controlled by cold
temperatures, and thus decrease in elevation southwards. The lower bounds
tend to be determined by the amount of growing season rainfall, which
increases to the south. Further south, the lower elevational boundary rises
again in response to higher evapotranspiration rates outweighing the
increased availability of summer rainfall. Transplant experiments of over 700
tree seedlings through the region showed that, where regional climate is not
stressful, the seedlings survived equally well on all aspects and elevations and
under different levels of cover. As temperature and rainfall constraints

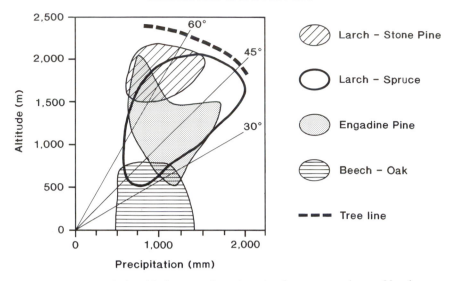

Figure 5.2 Relationship between four Austrian forest types, denoted by the
dominant or co-dominant tree species, and hygric continentality
Source: After Aulitsky *et al.* (1982)

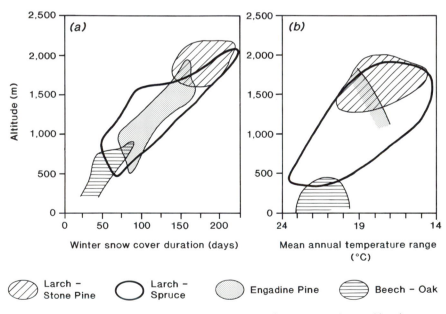

Figure 5.3 Relationship between four Austrian forest types, denoted by the
dominant or co-dominant tree species, and (a) winter snow-cover duration, (b) mean
annual temperature range
Source: After Aulitsky *et al.* (1982)

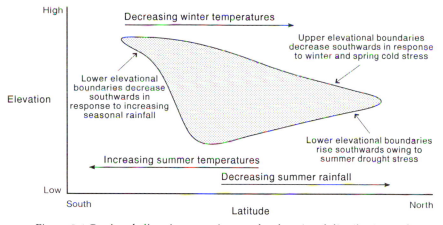

Figure 5.4 Regional climatic constraints on the elevational distribution and composition of plant communities along a regional climatic (latitudinal) gradient. As the gradients converge at the apex of the wedge, so the size of suitable habitat patches decreases and their diversity increases
Source: After Neilson and Wullstein (1983)

converged at the northern boundary of the species, seedling survival was progressively more constrained to specific slopes, elevations, and cover characteristics.

Animal species

Factors influencing animal species' distributions along elevational gradients are complex, and not solely climatic or even ecological. In some instances, evolutionary aspects of the species may be important. This appears to be the case for fruit bats (Pteropodidae) and murid rodents (Muridae) in the Philippine island group (Heaney *et al.* 1989; Heaney and Rickart 1990). Among fruit bats, species richness is highest in primary lowland forest and falls with increasing elevation, local endemism is moderate to low, and speciation is associated with colonization across oceanic channels. On the other hand, among murids, species richness rises with increasing elevation from lowland forest to montane forest and then drops a little in upper mossy forest. In addition, local endemism in murids is very high, and most speciation is associated with changing montane habitats within single islands resulting from Pleistocene climatal and vegetational changes. None the less, ecological influences on altitudinal sequences of animals are strong and can give clear signs in the range and abundance of species. To illustrate this point, the example of altitudinal bird and rat distributions on Mauna Loa in the Hawaiian Islands will be considered.

The flora and fauna along an elevational transect on Mauna Loa, running

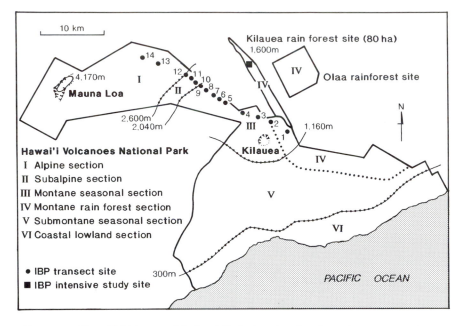

Figure 5.5 Hawai'i Volcanoes National Park showing the six environment sections, the fourteen Mauna Loa transect sampling sites, and the Kilauea rain forest site
Source: After Mueller-Dombois and Bridges (1981)

35 km from 3,660 m near the summit to 1,190 m near the summit of Kilauea, have been studied in great detail as part of the International Biological Programme (Figure 5.5) (Mueller-Dombois *et al.* 1981). The broad ecological conditions along the transect are summarized in Figure 5.6. For vegetational analysis, samples (relevés) were laid out nearly continuously along 25 km of the transect, from the slope base at 1,190 m to the upper alpine scrub limit at 3,050 m. Three relevés were placed in each of the twelve 2,150 m altitudinal intervals. Statistical analysis of the data revealed seven zones of woody plants (Figure 5.7). Zones I and II, the alpine and subalpine transect zones, extend through areas of lava-rock outcrop that contain only small amounts of fine soil material. They are occupied primarily by shrub species, though the tree species *Metrosideros collina* subspecies *polymorpha* appears in Zone II. Zones III and IV, mountain parkland and savanna, are underlain by lava covered by a blanket of volcanic ash that thickens and becomes more continuous downslope. In these zones, shrub-species diversity decreases, tree-species diversity increases, and herbaceous life-forms, especially grasses, become more abundant. In Zone V, *Metrosideros* dry forest, most of the shrubs, except those confined to high altitudes (*Vaccinium peleanum* and *Styphelia douglasii*), reappear while the other tree species disappear. In Zones VI and

110

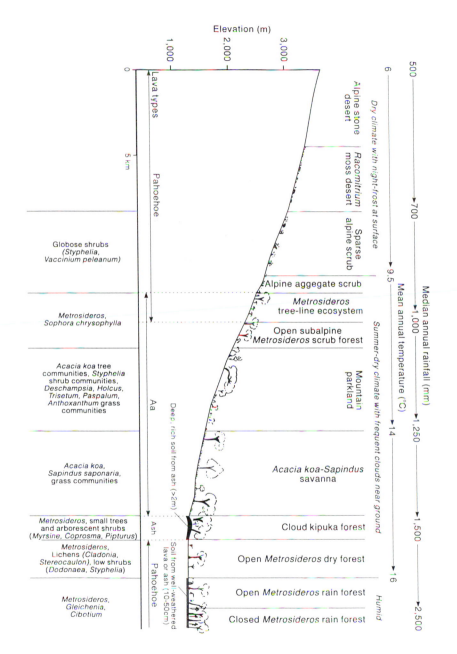

Figure 5.6 Profile diagram of the Mauna Loa transect showing vegetation, climate, and substrate

Source: After Mueller-Dombois and Bridges (1981)

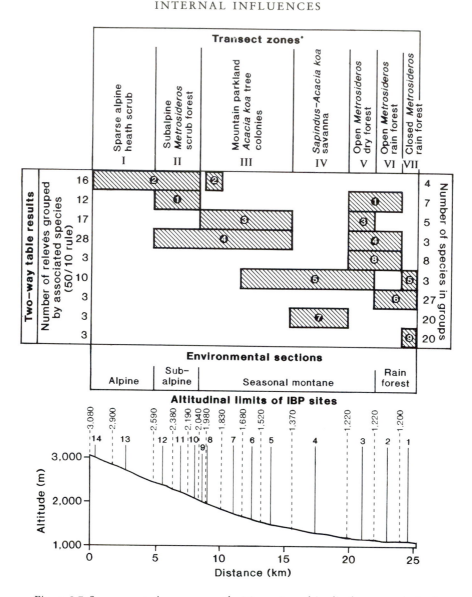

Figure 5.7 Seven vegetation zones on the Mauna Loa altitudinal transect derived from dendrograph and two-way table techniques for forty-eight plant community samples

Source: After Mueller-Dombois *et al.* (1981)

VII (open and closed *Metrosideros* rain forest), ferns become prevalent, shrubs associated with rock outcrops decrease in zone VII, and *Metrosideros* is dominant only on deeper soil in the rain forest.

Twenty-two bird species were recorded on the Mauna Loa transect

(Mueller-Dombois *et al.* 1981). These were grouped, using the Sørensen community coefficient based on presence–absence data, into six well-defined avian communities. These are indicated on Figure 5.8 as transect zones. The distribution along the transect of one indigenous (*Pluvialis dominica*) and eight endemic birds is portrayed in Figure 5.8. Four species groups were recognized. Group 1 has one member – the Hawaiian thrush or 'ōma'o (*Phaeornis obscurus obscurus*). This is the only bird living beyond the tree-line in the sparse alpine scrub. It also lives in the closed rain forest. Its distribution is therefore rather odd, occupying the ends of transect but not the intervening areas. Its absence from the middle part of the transect appears to result from competition with another frugivore, the exotic leiothrix (*Leiothrix lutea*). Group 2 comprises a spatially heterogeneous set of open area birds. The Hawaiian goose or nēnē (*Branta sandwichensis*) occurs from the tree-line throughout the subalpine forest and mountain parkland zones. The golden plover or kōlea (*Pluvialis dominica*), an indigenous bird, occurs mainly in the mountain parkland and open *Metrosideros* dry forest. It is absent from the savanna zone, probably because it dislikes the tall grass growing there. The Hawaiian owl or pueo (*Asio flammeus sandwichensis*) is confined to the mountain parkland and savanna zones. Group 3 comprises three endemic species ranging from the lower subalpine scrub to the rain forest. The Hawaiian hawk or 'io (*Buteo solitarius*) is found throughout the range but the distribution of both the Hawai'i 'elepaio (*Chasiempis sandwichensis sandwichensis*) and 'i'iwi (*Vestiaria coccinea*) is interrupted by the *Metrosideros* dry forest zone. The differences in distribution are attributable to differences in general behaviour and feeding habits: the 'elepaio and 'i'iwi favour colonies of *Acacia koa* trees or other forest groves with closed canopies, neither of which grow in the *Metrosideros* dry forest zone. Group 4 consists of the two most abundant native honeycreepers – the Hawai'i 'amakihi (*Loxops virens virens*) and the 'āpapane (*Himatione sanguinea*). The wide distribution of these species indicates broad tolerance of temperature and rainfall regimes. The slight fall in 'āpapane density in the *Metrosideros* dry forest suggests that it prefers closed tree canopies.

Several conclusions may be drawn from the elevational distribution of these (and exotic) birds on Mauna Loa, and from similar studies in the Peruvian Andes (Terborgh 1971; Terborgh and Weske 1975). First, distinct patterns of bird distribution are related to the distribution of major vegetation types along transects. Second, on Mauna Loa, the pattern of bird distribution is correlated to environmental factors, including vegetation. This can be seen in the upper limits of *Himatione* and *Loxops* that end abruptly at the tree-line. Above the tree-line, no individuals were seen, but just below the tree-line they live at high densities. *Vestiaria* and *Chasiempsis* exhibit sharp density changes where savanna vegetation changes to open *Metrosideros* dry forest. Third, vegetation structure exerts such a profound influence upon bird species' distributions that the distributional effect of competition

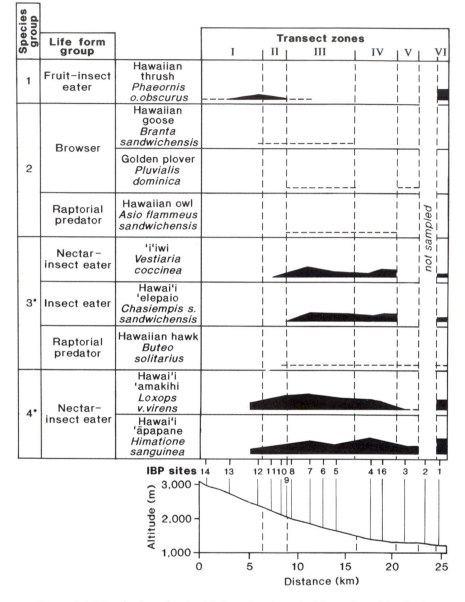

Figure 5.8 Distribution of native bird species along the Mauna Loa altitudinal transect. The six transect zones were established using the Sørensen index. The black areas for each species shows abundance (birds per 40 ha), the maximum abundance being 7 units (for example, *Himatione sanguinea* at site 4): 7 units = 160 to 300 birds; 6 units = 120 to 160 birds, 5 units = 70 to 120 birds, 4 units = 30 to 70 birds, 3 units = 15 to 30 birds, 2 units = 5 to 15 birds, and 1 unit = 2 to 5 birds; the dashed line represents one bird per 40 ha. Numbers with asterisks are computer-generated species groups
Source: After Mueller-Dombois *et al.* (1981)

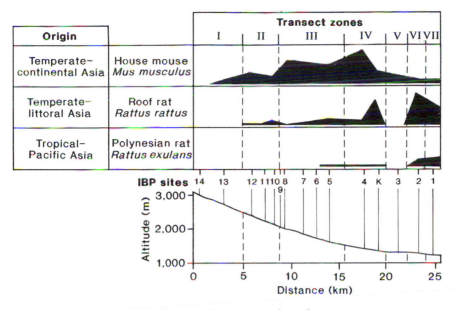

Figure 5.9 Distribution of three rodent species along the Mauna Loa transect. The black areas for each species is abundance in equal units of five animals trapped per year, with a maximum of 9 units = 41 to 45 animals (e.g. *Mus musculus* at site 4); 8 units = 36 to 40 animals, 7 units = 31 to 35 animals, 6 units = 26 to 30 animals, and so on. The horizontal line = one animal. K is the closed kipuka forest in the savanna zone IV

Source: After Mueller-Dombois *et al.* (1981)

between species is of little import. This contrasts with a similar altitudinal gradient in the Peruvian Andes where competition plays a significant role (Terborgh and Weske 1975). The difference between the two transects may reflect the contrasting settings – Mauna Loa an island setting with low avian diversity; the Peruvian Andes a continental setting with high avian diversity.

Four species of commensal rodents (Muridae), all introduced by humans, were recorded along the Mauna Loa transect. The Polynesian rat (*Rattus exulans*) arrived with Polynesian settlers. The roof rat (*Rattus rattus*), Norwegian rat (*Rattus norvegicus*), and the house mouse (*Mus musculus*) are all derived from Eurasian and American stock that were carried on ships to the Hawaiian Islands some time after their discovery in 1778. The Norwegian rat tends to live in close association with humans and was not recorded on the Mauna Loa transect. The other three species overlap broadly in their altitudinal ranges and do not form a distinct spatial group (Figure 5.9). Most abundant is the common house mouse which ranges from the alpine zone to the closed rain forest (and on down to sea-level). It is most abundant in the middle of the transect which carries closed herbaceous or grass vegetation with scattered trees (savanna zone) or tree groups (mountain parkland). Upslope and downslope of these zones the abundance of the house mouse declines. Downslope this decline occurs at the lower end of the savanna zone and is associated with the transition to closed forest. Upslope it is associated with a change from mountain parkland with closed grass cover patches, to open subalpine scrub forest with almost no herbaceous undergrowth. It is the only rodent to occupy the sparse alpine scrub zone where it builds secure nests in the pahoehoe lavas and survives on a meagre supply of *Vaccinium* and *Styphelia* fruits and insects. Clearly, the ecological optimum of the species is in vegetation with sunny grassland habitats. The roof rat was not trapped beyond the tree-line. Its peak abundance lies in closed forest vegetation and in open forest with dense fern undergrowth, and it was not observed in the *Metrosideros* dry forest zone. Thus, although the house mouse and roof rat have overlapping distributions along the altitudinal transect, their ecological optima do not coincide. The Polynesian rat shows a distribution limited by altitude, being rarely found above the rain forest.

COMMUNITIES AND ALTITUDE

Carl Linnaeus was aware that, on ascending mountainsides, a traveller would pass though different zones of life. The influence of altitude on the physical environment was noticed by Alexander von Humboldt. During his travels, Humboldt surveyed the physical, meteorological, and geological character-istics of equatorial mountains. These physical factors he correlated with the changes of vegetation that arose with increasing elevation (e.g. Humboldt 1817). Here then was the first detailed recognition of what later were to be called orobiomes. Today, change in species composition with increasing altitude is the usual focus of interest. Humboldt was chiefly concerned with the physiognomy of vegetation – the changing overall appearance of vegetation on a mountainside. He appreciated that the environments at different altitudes were favourable to different sorts of organisms, and, on equatorial mountains, he noticed an altitudinal succession of distinct communities of plants. Analogous successions of plant communities were found on other mountains, and in the groups of plants that occupied different latitudinal belts between the equator and the poles. Thus Humboldt believed that he saw on the slopes of Chimborazo, in the equatorial Andes, a full series of floral belts typical of the tropics, of the temperate regions, and of the boreal or Arctic zone. Tree-ferns and palms thrived at the mountain's base. Broad-leaved forests grew on higher slopes, and gave way to deciduous woodlands, often dominated by birch. The deciduous forests were replaced in the next higher elevational zone by coniferous forests, which in turn were superseded at greater altitudes by grassland and dwarf shrubs. And lastly, atop the mountain, Humboldt expected to find bare rock and scree with permanent snow if the peak stood high enough. To Humboldt, therefore, a mountain was a miniature version of a hemisphere whereon the floral zones of the Earth were duplicated on a vertical scale. Similarly, Clinton Hart Merriam (1894), after exploring the San Francisco Mountains, northern Arizona, in the late 1880s, suggested that zones of vegetation found on the flanks of mountains corresponded to the zones of vegetation found within latitudinal belts. He estimated that each mile of altitude was equivalent to 800 miles of latitude.

Altitudinal zones

During the twentieth century, the vegetational zoning on most of the world's mountains has been mapped. The basic zones encountered on ascending a mountain are submontane, montane, subalpine, alpine, and nival. A few examples will illustrate the chief changes related to altitude.

First, the vegetational response to altitude in Venezuela will be described (Walter 1973: 34). Venezuela lies between the equator and latitude 12° N. It

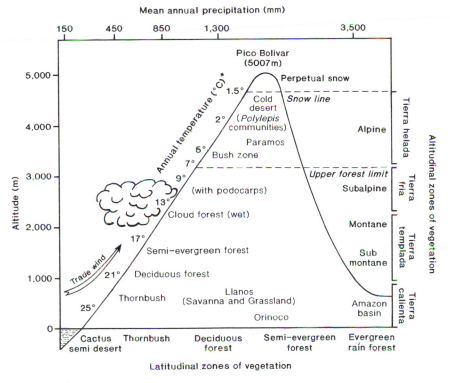

Figure 5.10 Altitudinal and latitudinal vegetation zones in Venezuela
Source: After Walter (1973)

117

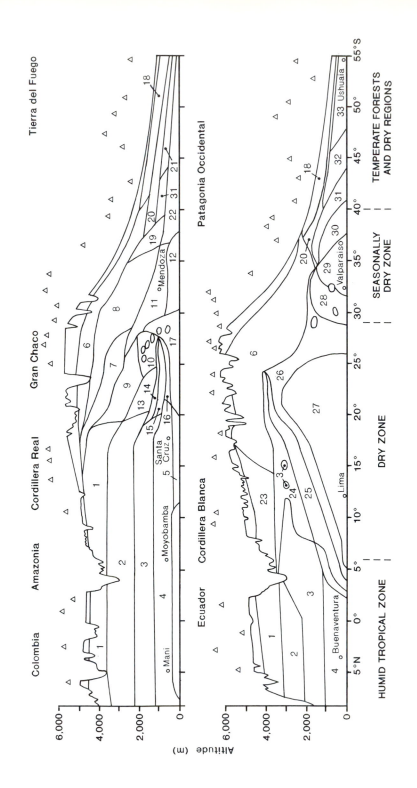

contains all altitudinal zones, from sea-level to the snow-capped Pico Bolivar standing at 5,007 m. The northern part of the country is open to trade winds between November and March. These winds bring rain to the mountainous districts inland. The lowlands have a dry season lasting five months and a wet season lasting seven. In the far south, in the Amazon Basin, at least 200 mm of rain fall every month. There is thus a latitudinal gradient of annual rainfall. It ranges from 150 mm at coastal La Orchila, to 3,500 mm in the Amazon Basin. Superimposed on this climatic gradient is an elevational climatic gradient. In mountainous regions, precipitation increases rapidly on windward slopes up to the cloud level, then decreases again above this. Temperature declines simultaneously at a rate of 0.57°C per 100 m of altitude. The inner valleys of the Andes are located within a rain shadow and are exceedingly dry. The response of the vegetation to these latitudinal and elevational climatic gradients is summarized in Figure 5.10. At sea-level, the vegetational sequence from north to south (dry to wet) is: cactus semi-desert, thornbush, deciduous forest, semi-evergreen forest, evergreen rain forest. Elevational changes depend on latitude, as shown on the diagram. Notice that the conventional altitudinal zones are given Spanish names in South America. An important lesson from this example is that it is not elevation or latitude alone that determine the climate – aspect and degree of shelter from rain-bearing winds make a substantial difference.

The vegetation of the Andes as a whole clearly shows the effect of an elevational gradient superimposed upon a latitudinal gradient and local effects (Figure 5.11) (Ives 1992). The northern Andes lie in the humid tropics where the high humidity supports lush vegetation. Tropical rain forests and other kinds of evergreen and deciduous forests are dominant. Altitudinal belts of vegetation are approximately symmetrically disposed on eastern and

Figure 5.11 (opposite) Altitudinal and latitudinal vegetation belts from Colombia to Tierra del Fuego. The top section is through the eastern Andes and the bottom section is through the western cordilleras. Triangles signify mountain peaks. Vegetation belts are identified by numbers: 1 Microthermal tropical mountain grasslands (paramo). 2 Mountain cloud forest. 3 Tropical montane forest. 4 Lowlands wet tropical forest. 5 Tropical moist savanna. 6 High mountain vegetation (puna). 7 Mountain pastures. 8 Mountain steppe. 9 Tall grassland. 10 Dry prairie. 11 Patchy woodland. 12 Subtropical thorn steppe. 13 Subtropical aliso forest. 14 Subtropical mirtaceas forest. 15 Subtropical laurel forest. 16 Subtropical pacara forest ecotone. 17 Subtropical thorn savanna. 18 Patagonian mountain bushland. 19 Mountain steppe with some timber. 20 Mixed subantarctic forest. 21 Subantarctic forest. 22 Patagonian steppe and bushland. 23 Moist puna belt. 24 Mesophyte bush vegetation. 25 Succulent vegetation. 26 Low thorn bush vegetation. 27 Peruvian desert and semi-desert with loma vegetation. 28 Thorn bush and succulent vegetation. 29 Wet temperate Chilean hardwood forest. 30 Deciduous forest. 31 Valdivian forest. 32 Patagonian inland forest. 33 Subantarctic evergreen forest
Source: After Ives (1992)

western flanks. The sequence runs from lowland wet tropical forest, through tropical montane forest, cooler mountain cloud-forest (variously termed tierra fria, selva nublada, Nebelwald, and moss forest) consisting of dense stands of ferns, epiphytes, and bamboo, to microthermal tropical grassland (paramo) above the tree-line. Beyond the snow-line lies the nival zone. The central Andes is a dry zone and the aridity influences the vegetation. On the coastal plain, the aridity is greatest in the Atacama desert, one of the driest places on Earth. The arid zone supports desert vegetation: thornbush, succulents, and loma vegetation supported by the garua, a dense mist that drifts northwards along the Peruvian coast. At higher elevations, the great high plateau, or altiplano, is covered by puna vegetation. Three types of puna occur along the humidity gradient running north to south. The northerly moist or humid puna crosses the Andes as a swath running from north-west to south-east. It consists of an uninterrupted mat of grassy plants. The dry puna is dominated by bunch grasses with resinous evergreen shrubs and cushion plants. Desert puna is characterized by xerophytic plants such as cactus and thorny shrubs. The eastern flank of the central Andes has a similar set of vegetation belts as those in the humid tropical zone, though the cloud forest is in this central region is called ceja de la montana or 'eyebrow of the mountain'. The southern Andes comprise a seasonally dry zone and a temperate zone. Moving south, eastern slopes become drier and western slopes become wetter. Along the coast immediately south of the Atacama Desert is an area of thorn scrub. This gives way to a sclerophyllous woodland and scrubland growing under a Mediterranean climate, similar to analogous formations growing elsewhere in the world but dotted with tall-growing trees. At higher elevations an extensive belt of desert puna gives way southwards to zones of mountain grassland and mixed forests. Comparable latitudes on the eastern flanks become progressively drier than on the west in moving south. Consequently, the vegetation becomes scrubbier and more xerophytic. The southern tip of the Andes, south of Valvivia, carries *Nothofagus* forest and other types of cool temperate forest on the west slopes.

An interesting study of soil and vegetation sequences along an elevation gradient was made in the San Jacinto Mountains and part of the Santa Rosa Range in southern California, about 120 km east of Los Angeles (Hanawalt and Whittaker 1976). Soils and vegetation were sampled in eleven ecosystem types on the inland slopes. The elevational gradient covered 3,000 m. Over the gradient, mean air temperature declined at rate of about 8.2°C per 1,000 m, and mean annual precipitation ranged from 7.6 cm in the lower deserts to 112–124 cm in the uppermost forest. All the soils were well drained and formed in granite or material derived from granite, so that variations in soil properties should have reflected climatic influences directly or by way of vegetation. Sites were chosen to represent typical vegetation with minimal disturbance on moderate to gentle slopes. The main changes in vegetation

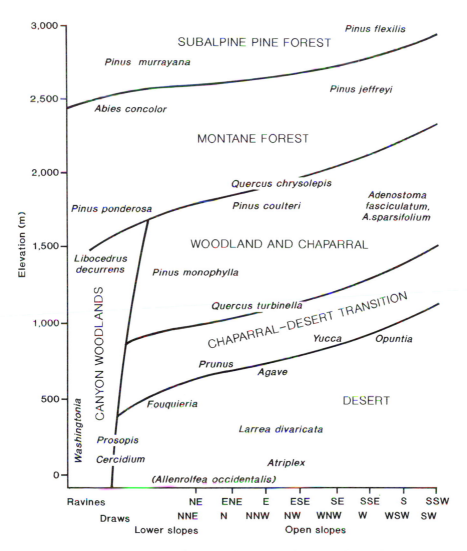

ALTITUDE

Figure 5.12 Vegetation in relation to elevation and topography on the inland slope of the San Jacinto and Santa Rosa Mountains, California. Mean position of chief vegetation types, based on 200 samples, are outlined. Although boundaries between vegetation types are drawn on the diagram, the variation is regarded as a continuum of intergrading types. Only the 'climax' types are shown; the vegetation in the field is extensively disturbed by fire, especially at high elevations. The pattern is generalized to include the two adjacent ranges, but there are differences between them: *Pinus murrayana* forests occur in the San Jacintos, and the dry phases of the chaparral are dominated by *Adenostoma fasciculatum* in the San Jacintos but by *Adenostoma sparsifolia* in the Santa Rosas
Source: After Hanawalt and Whittaker (1976)

121

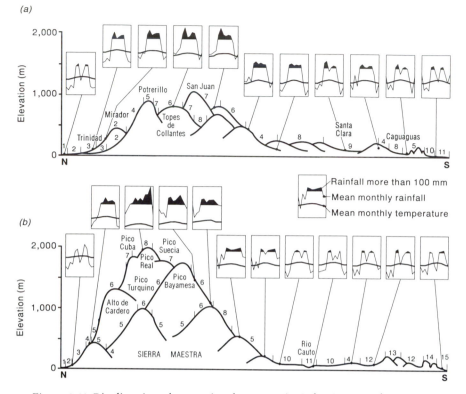

Figure 5.13 Bioclimatic and vegetational transects in Cuba. (a) Escambray Range, Las Villas Province. (b) Western Sierra Maestra Range, Oriente Province. Vegetation types in (a) are: 1 Sandy seashore. 2 Dry limestone forest. 3 Limestone thorn scrub-woodland. 4 Semi-deciduous forest. 5 Mogote formation. 6 Seasonal evergreen forest. 7 Montane rain forest. 8 *Roystonea* grassland. 9 Lowland serpentine scrub woodland. 10 Alluvial or swamp forest. 11 Mangrove forest. Vegetation types in (b) are: 1 Littoral rock pavement. 2 Littoral thorn scrub. 3 Dry limestone forest. 4 Semi-deciduous forest. 5 Seasonal evergreen forest. 6 Montane rain forest. 7 Elfin mossy forest. 8 Elfin woodland and thicket. 9 Montane pine forest. 10 *Roystonea* grassland. 11 Alluvial forest. 12 Low palm serpentine grassland. 13 Lowland serpentine scrub woodland. 14 Mogote complex. 15 Mangrove forest
Source: After Borhidi (1991)

followed the altitudinal changes in climate, though topographic position exerted an influence (Figure 5.12). Major changes in stature and structure, percentage of ground covered, and species composition were observed along the elevational gradient. Shrub coverage had two modes, one in the moist soils of the saline desert, and a second in the pine–oak woodland. The area of low coverage between the two modes corresponded to the dry deserts. Above the pine–oak woodland, shrub cover decreased as trees came to dominate the forests, though above the chaparral individual shrubs tended to become larger. Tree cover and height increased from the piñon–juniper woodland, through the Coulter pine–oak woodland, to the ponderosa pine–oak woodland, and then declined towards the highest elevations.

Altitudinal changes in Cuban vegetation are well documented (Borhidi 1991). Two bioclimatic and vegetational transects are depicted in Figure 5.13. The first is a transect through the Escambray Provinces in the central region. The northern ranges of Villa Clara, the serpentine hills of Santa Clara, and the northern foothills of the Sierra de Escambray experience a seasonal climate with dry winters. The vegetation in this region is influenced by the mosaic-like pattern of the underlying rock. Limestone supports semi-deciduous forest, serpentine dry evergreen shrubland. The northern slopes of the Escambray mountains are wetter than the northern foothills and the dry season shorter. They support seasonal evergreen forest. Only on the highest parts of Escambray and on northern slopes and valleys does the dry season disappear and montane rain forest flourish. Southwards from the central plateau (Topes de Collantes), rainfall amounts fall rapidly and a dry winter season becomes established. Belts of semi-deciduous forest, evergreen shrub forest, and dry evergreen forest are found on descending these southern slopes. In the southern lowlands of the mountains, the climate is dry for eight to nine months each year and sclerophyllous and thorny evergreen thickets are found. In the western part of the Sierra Maestra Range, in Oriente Province, a similar pattern of vegetation is established, though the belts are wider. The colline regions of Oriente region are very dry. Only in the foothills of the Sierra Maestra is the dry season appreciably reduced in length. The chief difference between the two regions is that the vertical precipitation gradient is much steeper in the Sierra Maestra Range, and this eliminates a dry season. It is probably for this reason that seasonal rain forests cover the foothills and the montane rain-forest belt reaches lower altitudes than in the Sierra de Escambray. Upwards of 1,400 m, above the montane rain-forest belt, a mild and wet temperate climate occurs supporting elfin mossy forest (or cloud forest). Above 1,800 m, a cooler temperate climate supports a belt of elfin thicket.

The zoning of animal communities on mountains is not strongly marked, but altitude undoubtedly has an influence on animal species diversity. This is seen in R. J. Ranjit Daniels's (1992) study of amphibian ranges and distributional patterns on the Western Ghats, part of the Malabar biogeographical region in south India. The Western Ghats run unbroken for 1,600 km between latitudes 8° and 21° N, save for the 30-km-wide Palghat Gap at around 11° N. They are very rich in amphibian species (frogs, toads, and caecilians), containing 117 species, 89 of which are endemic. Climatic effects on amphibian species diversity are expressed through altitudinal and latitudinal gradients. Species diversity is greater in the southern half of the Western Ghats, south of 13° N, and in hills of low-to-medium elevation, than it is in the northern half of the area and in high hills. This suggests that the presence of high hills does not add to amphibian species diversity in the Western Ghats, a finding contradicting the prediction made by Robert Inger and his colleagues (1987). The differences in diversity appear to relate to more widespread rainfall and less climatic variability in the southern area. Unlike birds and flowering plants in the area, where endemics are found chiefly on the higher hills, endemic amphibians are confined largely to the lower altitudinal range, with the majority in the range 800 to 1,000 m.

Altitudinal vegetation zones: discrete communities or continua?

The above examples, and many more, affirm that vegetation on mountains is zoned. Nevertheless, an interesting question may be posed about this zoning: are altitudinal vegetation belts discrete communities, or are they simply expressions of a continuous variation in plant species with increasing altitude?

Michael Auerbach and Avi Shmida (1993) sought to see if communities are discrete units or whether they are merely abstractions within a vegetational continuum by looking at altitudinal vegetation change on Mt Hermon, Israel. If plant communities are discrete entities, a concordance in species' geographical ranges within a particular community would be expected to occur. Mt Hermon is nicely suited to studying altitudinal vegetation gradients because it has a fairly constant slope and relatively homogeneous geomorphological and edaphic conditions. Previous work established three vegetation zones on the mountain. The low zone, extending from 300 to 1,300 m, is evergreen Mediterranean maquis characterized by evergreen sclerophyllous shrubs with an evergreen oak (*Quercus calliprinos*) as the dominant tree below 900 m, and with a deciduous oak, *Quercus boissieri*, and *Quercus calliprinos* as co-dominants above 900 m. The middle zone, extending from 1,300 to 1,900 m, is xero-montane open forest. This consists of widely dispersed and stunted deciduous trees (up to 2 m tall), mainly oaks and members of several genera of the Rosaceae, with perennial herbaceous vegetation, chiefly grasses and spiny plants, in between. The high zone, extending from 1,900 to 2,800 m, is a subalpine tragacanthic belt comprising sparse, cushion-like shrubs, short herbaceous species, and geophytes typical of most arid alpine regions in the Middle East.

To establish whether these zones are discrete communities, the altitudinal distributions of all 903 vascular plant species on Mt Hermon were culled from previous studies, including 1,270 10 × 10 m relevés taken along the altitudinal gradient and 8,700 herbarium sheets collated expressly for documenting plant ranges along mountainsides. The flora on Mt Hermon was simulated a thousand times using 100-m-wide altitudinal bands, and the expected number of species starting and ending their altitudinal ranges in each of the twenty-six 100-m altitudinal bands was computed as the mean of the 1,000 simulations ranges. The procedure was repeated using a 50-m-wide altitudinal band to test the effect of sampling scale. By comparing expected and observed species' endpoints, it was found that there was no significant difference between the cumulative frequency distributions of observed and expected downslope boundaries, but the distribution of upslope boundaries differed significantly, with values greater than the critical value occurring at the 900–1,000-m and 1,400–2,200-m bands. Fewer species than expected ended their range at 900–1,099 m, and more than expected ended their range in the 1,400–2,299-m band. So, species starting low on the mountain tended to have longer ranges than expected. Similar results were obtained using the 50-m altitudinal bands. The analysis showed that the spatial arrangement of range boundaries differed from a random pattern at some altitudes, but despite this aggregation of boundaries there was little evidence for discrete zones or communities. If there are discrete zones on Mt Hermon, then an unexpectedly large number of upslope and downslope boundaries should coincide in the ecotones between the communities. Certainly, for both 100-m and 50-m sampling bands, upslope and downslope boundaries did coincide at 1,200 m and near the mountain top at 2,800 m. At 1,200 m there appears to be a swift transition from maquis dominated by *Quercus calliprinos* to an open deciduous forest characterized by *Quercus boissieri* and rosaceous shrubs. This transition does seem to be an ecotone. The second apparently abrupt transition occurring near the mountain top is an artefact of the low numbers of observed and expected downslope boundaries. Taken as a whole, the results suggest that discrete communities do not exist on Mt Hermon. On the other hand, given the aggregation of some species' borders, neither is the distribution of plant species perfectly continuous. It is possible that the montane species' distributions accord with another variant of the vegetation continuum concept (see Austin and Smith 1989), but this possibility was not explored. The conclusion reached was that 'in the absence of sharp edaphic discontinuities, vegetation change along altitudinal gradients generally appears to be a subtle, continuous process' (Auerbach and Shmida 1993: 32). A like conclusion was drawn from a similar study in Tasmania where, although three altitudinal vegetation zones were readily discerned in the field, there is enough continuity between the zones to invalidate ideas about precise altitudinal boundaries (Ogden and Powell 1979).

Tree-lines

Tree-lines, steep ecotones in which a noticeable change of dominant life-form occurs (Plate 5.2), are seen very clearly in the circumpolar boundary between taiga and tundra vegetation in the Northern Hemisphere, and in the boundary between subalpine vegetation and low-growing alpine vegetation on mountains. They can tell us much about elevational influences on plants. In particular, they seem like a good place to look for discontinuities in communities, as they are defined by trees growing at their limit of climatic tolerance.

Plate 5.2 The tree-line on slopes flanking the Peyto glacier outflow, leading into Peyto Lake, Rocky Mountains, Alberta, Canada. As well as being stressed by climatic factors, the trees in this environment have to tolerate frequent debris avalanches.
Photograph by David N. Collins

Plate 5.3 The tree-line in the Southern Alps, New Zealand. The photograph is taken looking from the subalpine zone down to the southern beech (*Nothofagus*) forest and braided Waimakariri River.
Photograph by Brian S. Kear

Ecologists generally distinguish three types of altitudinal tree-line, though the terminology varies (see Wardle 1974; Hustich 1979). The Waldgrenze, or forest line (referred to as the timberline in the United States), is the upper limit of tall, erect trees growing at normal forest densities. Above this is the Baumgrenze, or tree-line, the limit at which individuals recognizable as trees (sometimes defined as more than 2 m high) grow. Higher again lies the tree-species line, the point up to which tree species will grow, albeit in a deformed form. In many cases, the tree-line and tree-species line coincide, as in the southern beech forests of New Zealand (Plate 5.3), and all three lines may coincide in some situations. Where this happens, the discontinuity between trees and low-growing vegetation is very sharp. Between the timberline and the tree-species line lies the band known as the Kampfzone. In English, Kampfzone means 'battle zone', or, perhaps a better rendition would be 'stress zone'. Trees live in the Kampfzone, but, owing to the severity of the environmental conditions, they have a deformed, dwarf, or prostrate appearance, all of which signalize stress. Trees in the Kampfzone are described as Krummholz, meaning 'crooked wood'. A common form of deformity is known as 'flagging': the crown of a 'flagged' tree grows mainly on the lee side of the trunk so that the tree resembles a flag flying in the wind. In Austria, in the high core of the central Alps, such as around Mount

Patscherkofel (2,247 m), the Waldgrenze runs at about 1,950 m (Pears 1985: 135). Spruce (*Picea abies*) and larch (*Larix decidua*) grow robustly up to this height. The Baumgrenze lies at about 2,150 m and is marked by the last appearance of the stately Arolla pine (*Pinus cembra*). Dwarf specimens of spruce, larch, and Arolla pine are found in the Kampfzone. A word of caution is necessary here – some trees adopt dwarf forms whatever the environmental conditions. In the central Alps, the dwarf mountain pine (*Pinus mugo*) thrives in the Kampfzone, but retains its dwarf habit even when grown at lower altitudes. Its Krummholz character is therefore genetically determined and not the outcome of environmental stresses.

Tree-lines are not always sharp. In the Caucasus Mountains, for instance, they may be sharp or diffuse (Armand 1992). On the southern slope of Assara Mountain, the transition from forest to alpine meadow has two steps: the shrub belt, in which the hazel (*Corylus avellana*) is the dominant species,

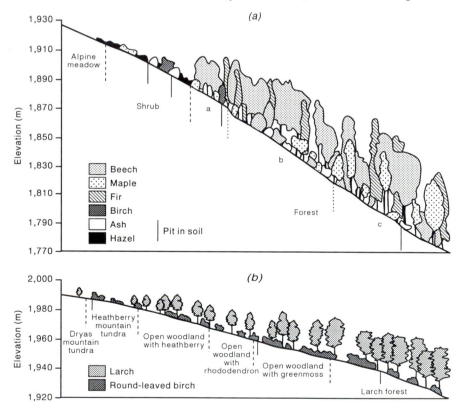

Figure 5.14 Gradual and abrupt tree-lines. (a) Upper boundary of the fir-beech forest on the southern slope of Assara Mountain in the west Caucasus Range. The letters a, b, and c denote forest subzones. (b) Upper boundary of the larch forest on the north-eastern slope of the Balakhtyn-Shele Range, East Sayans
Source: After Armand (1992)

lies above the forest line but within 50 m it is replaced by alpine meadow (Figure 5.14a). On a north-eastern slope of the Balakhtyn-Shele Range, in the East Sayans, the forest line is diffuse (Figure 5.14b). Larch (*Larix sibirica*) forest, with occasional cedar trees (*Pinus sibirica*) and a subcanopy of round-leaved birch (*Betula rotundifolia*), gives way through about 100 m of elevation to mountain tundra with dryad (*Dryas oxiodonta*), blueberry (*Vaccinium uliginosum*), heathberry (*Empetrum nigrum*), and others. The results of the study show that, in these two mountains, a gradual altitudinal change of climate causes in one case a sharp replacement of a tree formation with a herbaceous formation, but in the other case a gradual change. The sharp transition occurs as several steps, but each step is clearly visible in the field.

Inverted tree-lines occur in topographic lows where cold air commonly drains into, and fills, mountain valleys to create frost hollows. This phenomenon is not very common but has been reported from Australia and the United States. On Mount Buffalo, Victoria, Australia, frost hollows occurring at about 1,520 m prevent the growth of snow gums (*Eucalyptus pauciflora*) and allow herbs, grasses, or dwarf-shrub vegetation to flourish. In Nevada, temperature inversions in the piñon–juniper zone (*Pinus monophylla–Juniperus osteosperma*) lead to a lower tree-line at about 1,470 m and treeless valley bottoms (Billings 1954). The upper tree-line stands at about 2,067 m which means that ridges are treeless. The belt covering the valley-side slopes between the lower and upper tree-lines, known as the thermal zone, favours the growth of piñon pine and juniper.

The causes of tree-lines are much debated. Some tree-lines are demonstrably determined by non-climatic factors. In the Chilean Andes, massive slope disturbances, chiefly the result of volcanic activity, impose severe constraints on the altitude of forest growth, as well as forest composition and structure (Veblen *et al.* 1977). In the Caucasus Mountains, Aleksey D. Armand (1992) attributes vegetational changes across the tree-line to biotic interactions, rather than external influences. He believes that competing communities affect each other not directly, but through modification of the secondary environment. The dominant species in the forest alter their environment to suit themselves. Likewise, alpine meadow species change their environment in such a way that the growth of forest trees is impaired. The different communities are thus self-reinforcing. The sharpness of the transition zone is then a reflection of the extent to which the communities change their internal environment: the more radical the environmental alteration, the sharper the ecotone.

Despite the cases mentioned above, most ecologists believe that climate is the root cause of tree-lines. Climate influences tree-lines through the carbon balance and mortality of tree species (Slayter and Noble 1992). Trees are the ideal life-form for maintaining a positive carbon balance when competition for solar radiation is important. As the tree limit is approached, other factors assume greater significance and the response of trees is to reduce the amount of carbon allocated to wood production and reduce growth height. Smaller, less woody, shrub-like trees, often with many stems, appear that are less susceptible to wind and snow damage. In environments where a tree fails to maintain a positive carbon balance for long enough, it will not survive. So, explanations of tree-lines require an understanding of plant ecophysiology at high elevations, and particularly that suite of environmental factors, mainly climatic but also edaphic, that influence photosynthesis (e.g. Tranquillini 1979; W. K. Smith and Knapp 1990). A strong influence of climate on tree-lines is suggested by the fact that the height of the tree-line is affected by the size of mountains and by latitude. Tree-lines in the massive mountains occupying the siliceous core of the central Alps are found at higher elevations than on the less massive, calcareous mountains (the South Tyrol and the Northern Alps) at roughly the same latitude. This difference may be attributed to climatic differences engendered by the Massenerhebung Effect. The same effect also occurs in more isolated mountain systems where the tree-line is lower than normal for the latitude. Furthermore, the height of tree-lines is modulated by latitude, implying a climatic control (Figure 5.15). As a rule of thumb, the height at which most tree-lines occur is marked by the 10 °C isotherm for the warmest month (Daubenmire 1954). The difference between tree-line height on continents and on islands may be attributed to the effects of continentality (Baumgartner 1980): mountains on continents experience more extreme ranges of seasonal temperatures than do mountains on islands, as well as smaller overall temperature lapse-rates. For a given latitude, the 10 °C isotherm will sit at higher elevations on continental mountain ranges than on insular ones. However, the lower tree-lines on islands may result in part from a more restricted gene pool, and in part from their maritime character that restricts the capacity of plants to adapt to cold environments.

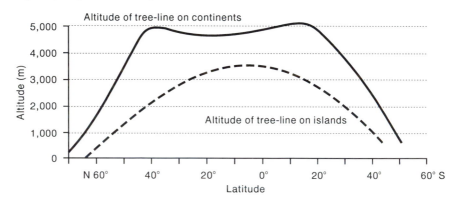

Figure 5.15 The relationship between the altitude of tree-lines and latitude on continents and on islands
Source: After Baumgartner (1980)

ALTITUDINAL ZONATION OF SOILS

Altitudinal trends

Many climosequences of soils have been established in mountainous terrain. Here, temperature decreases and rainfall increases with increasing elevation, often within short distances. The effects of these climatic changes on soil properties have been reported from a variety of areas. Such studies show that increasing elevation is associated with an increasing content of organic matter and nitrogen, an increasing ratio of carbon to nitrogen, and increasing acidity; and with a decreasing content of calcium, magnesium, and potassium. These elevational climosequences are influenced by latitude. For instance, forested soils gain humus with increasing altitude, chiefly because the drop in temperature leads to a slowing down in the rate of humus decomposition. The nearer a mountain is to the equator, the faster is the rate of humus gain: to double the amount of soil nitrogen in the Sierra Nevada, an ascent of 2,760 m is required; in India, the same gain is attained in an ascent of 1,350 m; and, in Colombia, the figure is just 890 m. These increases in altitude are equivalent to mean annual temperature falls of 14.6, 7.6, and 5.0 °C, respectively.

The type of humus produced depends largely upon latitude and parent material (Duchaufour 1982). In the temperate region, the middle and lower montane zone supports mull humus and acid brown soils. In the upper montane zone, biological activity is still sufficient to secure the formation of mull or thick mull-moder, providing some deciduous trees are mixed in with the conifers. This zone encourages the genesis of humic ochric brown soils. Above the montane zone, in the subalpine zone, the vegetation is dominated by species, such as conifers and Ericaceae, that promote acidification. Soils evolved in this more acidic environment are brown podzolics and podzols. On the alpine meadows, which lie above the tree-line, humus is a relatively active moder or mull-moder, but the subsoil is not well developed. Rankers form on silicate rocks and humic lithocalcic soils on limestones. This climosequence is best expressed on north-facing slopes. On south-facing slopes, the extra intensity of solar radiation with altitude offsets the lowering of temperature. The result is that the speed of organic matter decomposition is increased and the climatic belts become blurred. In the tropical zone, as in the temperate zone, belts of different vegetation types girdle the mountains. There are two big differences between the zones on temperate and tropical mountains. First, on tropical mountains there are no seasons. Second, organic matter accumulation, almost negligible in tropical lowlands, rapidly assumes importance with increasing altitude. In very wet conditions on granite, a peaty or hydromor horizon readily forms at the surface. The effect on soil formation is that surface podzolization is superimposed on ferrallitization to create an altitudinal humic ferrisol. These soils are associated with mountain forest with tree ferns. Nearer the summit, soil profiles become thinner and are capped by a very acid peaty humus. Three zones occur: the first zone contains humic ferrisols with secondary gibbsite in which hygrophilic shrubby forest grows; the second zone consists of humic cryptopodzolic soils supporting subalpine shrubs; and finally comes the zone of peaty alpine rankers that supports paramo.

Altitudinal climosequences

Altitudinal climatic gradients are normally more compressed than latitudinal and longitudinal gradients. In some ways, this climatic compression makes it easier to investigate 'vertical' climosequences than the more extensive 'horizontal' ones. Several altitudinal climosequences have been scrutinized within the last decade. One study investigated three soils developed in volcanic ash in Papua New Guinea's Enga Province (Chartres and Pain 1984). The three soils were sampled at altitudes of 1,040 m, 1,720 m, and 2,350 m, the sample points being located within about 40 km of each other. With increasing altitude, greater proportions of silt-sized, unweathered and partially altered primary minerals, and increasing molar ratios of calcium, magnesium, sodium, and potassium to aluminium, were discerned. These changes in soil properties were attributed to the temperature decline, and associated changes in evaporation and leaching, with altitude. An altitudinal climosequence on Volcán Barva, Costa Rica, looked at soils from six primary forest plots at altitudes of 100 m, 1,000 m, 1,500 m, 2,000 m, and 2,600 m (Grieve *et al.* 1990). All the soils were evolving in volcanic parent materials. Weathering intensity and organic matter decomposition were found to decrease with increasing elevation.

Earl B. Alexander and his colleagues (1993) carried out an investigation of soil–elevation relationships in South Cascade Range, northern California. The soils studied, all well drained, are present in the Tuscan Volcanic Plateau of late Cenozoic age lying west-southwest of Lassen Peak. Parent materials are andesitic and basaltic lava and mudflows (lahars). Three potential complicating factors in identifying and interpreting elevational transects were considered. The first complication was climatic change; this appears not to have been important in the study area. A second complication was changes of elevation caused by uplift. Again, these changes seem to have been unimportant in the study area. However, the third complication – the time factor of soil formation – did appear to exert an influence on soil evolution in the region. It was found that, on slopes up to 30 per cent, soil evolution reflects position in an altitudinal sequence rather than age, while on slopes exceeding 30 per cent, the length of soil evolution seems to be the predominant influence upon pedogenesis. For this reason, data from soils on slopes steeper than 30 per cent were disregarded. In total, 27 pedons were sampled at altitudes ranging from 270 to 2,030 m (Figure 5.16). The dominant soils and vegetation in the elevational sequence are shown in Figure 5.17.

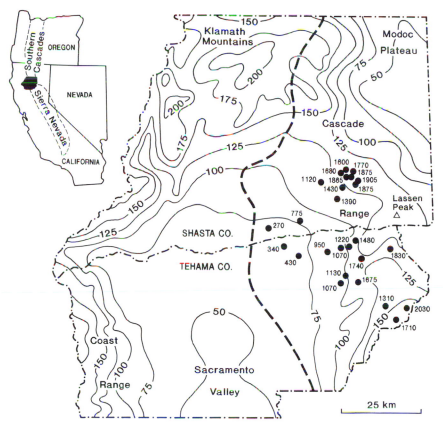

Figure 5.16 Location of pedons (dots) in an elevational climosequence superimposed on an isohyetal map (cm/yr) of Shasta and Tehama counties, California. Numbers by pedons are elevations (m). The dot–dash line demarcates the western limit of the Cascade Range volcanic rocks. There is no distinct boundary between the Cascade Range and the Modoc Plateau
Source: After Alexander *et al.* (1993)

Below 500 m, typic Rhodoxeralfs support blue-oak grassland with sparse digger pine trees. Above 500 m, warm, cool, cold, and very cold conifer forest is associated with mesic Palexeralfs, mesic Xeric Haplohumults, frigid Andic Haplohumults, and Haplocryands respectively. Black oak (*Quercus kelloggii*) is common in the conifer forest at lower altitudes. Soil morphology varies along the elevational sequence. Argillic horizons are best evolved in soil of the oak grassland. They occur in the lower coniferous forest but have more gradual upper boundaries than in the oak grassland and are thicker and have more clay. At higher elevations, the argillic horizon thins and contains less clay. No argillic horizons are found above 2,000 m. The redness of the argillic

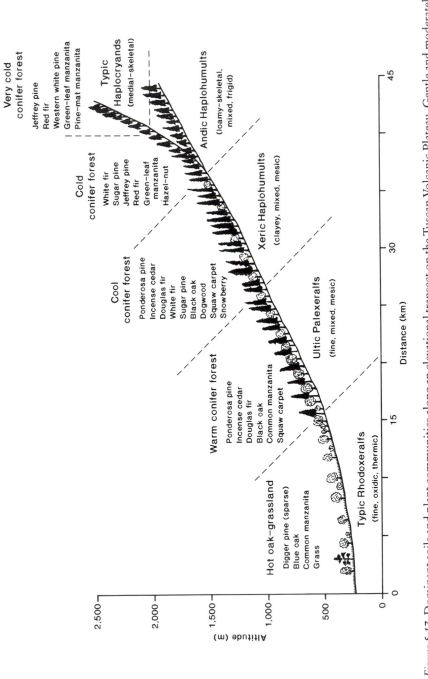

Figure 5.17 Dominant soils and plant communities along an elevational transect on the Tuscan Volcanic Plateau. Gentle and moderately steep slopes are depicted above 1,500 m. The shrub cover is sparse in all mature stands of natural vegetation, but it may be almost 100 per cent in disturbed areas

Source: After Alexander *et al.* (1993)

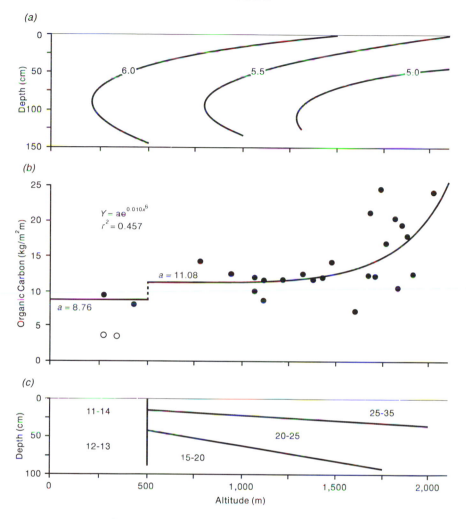

Figure 5.18 Soil properties along an elevational transect on the Tuscan Volcanic Plateau. (a) Soil pH versus depth and altitude. (b) Soil organic carbon (in upper metre of 'mineral' soil) versus altitude. (c) Ratio of organic carbon to total nitrogen (in 'mineral' soil) versus depth and altitude
Source: After Alexander *et al.* (1993)

horizon is greatest under the oak grassland, and fades with increasing altitude under the conifer forest. Soil properties vary in a regular way along the elevation gradient (Figure 5.18). The characteristic (but not necessarily predominant) clay minerals, from low to high altitude, are kaolinite, halloysite, hydrated halloysite, and allophane. Soil depth and clay content are greatest in the lower part of the conifer forest. Soil reaction is slightly acid throughout the solum in oak grassland (Figure 5.18a). At higher altitude, it is strongly acid in the surface layer, and very strongly acid in the subsoil. Soil organic carbon increases with altitude, the content in the upper parts of the conifer forest being about thrice that in the oak grassland (Figure 5.18b). The organic carbon–nitrogen ratio in surface horizons also increases with elevation, but the most abrupt increase is where grassland gives way to conifer forest (Figure 5.18c). Were not the discontinuity in carbon–nitrogen ratios at the grassland–forest transition so marked, the slight jump in soil organic carbon across the same transition zone would not have been detected. No apparent discontinuity exists for soil nitrogen. Piecing the information together, it would seem that soils with the deepest sola and highest clay contents have evolved in the lower conifer forest; and the reddest soils have evolved in the oak grassland. It is possible that erosion in the oak grassland has reduced the depth of solum and clay content of the soils found there.

SUMMARY

The plant communities girdling the Earth as broad zonal belts are paralleled in the plant communities encircling mountains: latitudinal and altitudinal climatic zones engender analogous communities. Individual animals and plant species often occur within a particular elevational band, largely owing to climatic limits of tolerance. However, as with latitudinal and longitudinal ranges of species, altitudinal ranges are influenced by a host of environmental factors that are best studied using a multivariate approach. This was exemplified by the study of tree species in the Rocky Mountains. Plant communities are subject to a variety of environmental influences, and not just to climatic ones. Much ecological work adopts a multivariate approach to vegetation sequences along elevational gradients. Debate surrounds a crucial question: are altitudinal communities discrete units, or are they continua? Current consensus favours the continuum concept, though discontinuities in vegetation sometimes do occur, as at the tree-line. Tree-lines themselves are not always purely the result of climatic constraints on tree growth, but involve pedological and biotic interactions as well. Soil climosequences following elevational gradients have been established. As with latitudinal and longitudinal climosequences, the convincing elevational climosequences involve rapidly responding soil properties such as carbon content, nitrogen content, and type of humus.

FURTHER READING

The author has detected a recent renewed interest in the effects of altitude on animals, plants, and soils. The best survey of mountain climates must surely be the book by Roger G. Barry, now in its second edition. John Gerrard's

Mountain Environments: An Examination of the Physical Geography of Mountains (1990) is well worth dipping into. Walter Tranquillini's *Physiological Ecology of the Alpine Timberline: Tree Existence at High Altitudes with Special Reference to the European Alps* (1979) provides a useful summary of ideas about tree-lines, but should be supplemented by more recent writings on the subject (e.g. Slayter and Noble 1992; Armand 1992; J. C. James *et al.* 1994). As for soils and altitude, the paper by Earl B. Alexander and his co-workers (1993) may be read with profit (see also Bäumler and Zech 1994).

6

SUBSTRATE

Geoecosystems are influenced by the substrate upon which they rest. Substrate may be defined as soil, rock debris, or rock, depending on the geoecosystem being considered: to animals and plants, the edaphosphere is the substrate; to the edaphosphere, the debrisphere is the substrate; and to the debrisphere, the lithosphere is the substrate. This chapter explores the influence of substrate on the pedosphere and biosphere.

ROCKS AND SOILS

Parent material

The earliest surveys and classification of soils, carried out during the nineteenth century, were often based upon the geology and composition of the material from which the superjacent soil had been formed. Early pedologists equated parent material with geological substratum. Soils were designated 'granite soils', 'glacial soils', and so forth, according to the type of geological material in which they were developed. They were also placed into two broad classes – residual soils and alluvial (or accumulated) soils. The family ties between rock and soil were made plainer in France and Germany: parent material was termed 'mother rock' (roche mère and Muttergestein). When soil-forming factors were formulated, parent material needed redefining more precisely. In the relatively shallow and 'youthful' soils of Europe, A and B horizons are fairly distinct from the underlying C horizon, the material from which the upper horizons are generally formed. In the United States, where deep and old soils are more common, parent material and C horizon advert to weathered rock or saprolite. The undecomposed rock at greater depth is labelled R, or G if it is gleyed material and K if it is carbonate-rich material. Considerable confusion surrounds the naming of soil horizons, though attempts are being made to rectify this messy situation (e.g. Bridges 1993).

There is some argument over what constitutes parent material. Is it exclusively material in the R horizon, or should it include material in the C horizon as well? If, as was suggested earlier in the book, soil is considered as rock that has come within the influence of the ecosphere, then all horizons will be part of the soil; and the true parent rock will occur below the depth that ecospheric influences can reach, that is, in the lithosphere. Material in the R horizon is likely to bear a few signs of weathering. So, technically speaking, it is not the unweathered state of the material in which the overlying soil has evolved. But normally, it is so like the true parent material that it can be regarded as such. A complication with the recognition of parent material is the additions and subtractions of rock by geoecological processes – volcanic ash, aeolian dust, eroded material – which may make a significant contribution in some soils (cf. Crocker 1952). These geoecological processes may lead to a situation where the parent material in which a soil evolves lies on a rock of very different composition, and which makes no contribution to pedogenesis. In these cases, an Arabic or Roman numeral precedes the letters denoting horizons evolved in material beneath the lithological discontinuity.

Parent material is indubitably a key factor in determining the course of soil genesis. At a local scale, the pattern of soil types reflects the pattern of parent materials: the distribution of soil series closely matches the distribution of the parent materials as depicted on drift geology maps. Parent materials are part of the debrisphere, the thickness and composition of which is an important determinant of edaphic properties and the distribution of animals and plants. The properties of the debrisphere are influenced by all environmental factors, including topography (acting through slope processes) and grandparent material. This is demonstrated by Yuichi Onda's (1992) study of regolith and

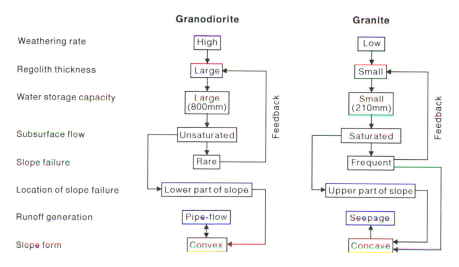

Figure 6.1 Schematic diagram of feedback in regolith-slope systems on granodiorite and granite in experimental drainage basins of the Obara area, central Japan
Source: After Onda (1992)

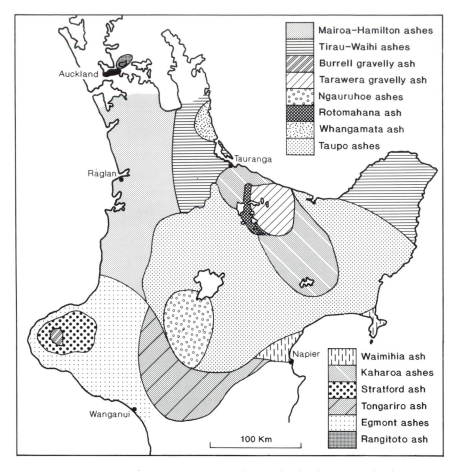

Mairoa–Hamilton ashes
Tirau–Waihi ashes
Burrell gravelly ash
Tarawera gravelly ash
Ngauruhoe ashes
Rotomahana ash
Whangamata ash
Taupo ashes

Waimihia ash
Kaharoa ashes
Stratford ash
Tongariro ash
Egmont ashes
Rangitoto ash

Auckland
Tauranga
Ráglan
Napier
Wanganui

100 Km

Figure 6.2 Volcanic parent materials, North Island, New Zealand
Source: After Gibbs (1980)

slope formation on two contrasting rock types – a medium-grained granodiorite and coarse-grained granite – in four experimental drainage basins of
the Obara area, central Japan. A feedback system exists among weathering
rate, hydrological processes, slope failures, regolith thickness, and slope form
that is very different on the two lithologies (Figure 6.1). On granodiorite, a
high rate of weathering produces a thick regolith with a large water-storage
capacity; water moves downslope as unsaturated throughflow and slope
failures are rare, which helps to maintain a thick regolith; such slope failures
as occur are located on the lower part of hillslopes promoting the evolution
of convex slope profiles. On granite, a low rate of weathering generates a thin
regolith with a small water-storage capacity; water moves downslope as

140

saturated throughflow and slope failures are common, which helps to keep the regolith thin; the frequent slope failures occur on the upper part of hillslopes promoting the evolution of concave slope profiles.

Many rocks produce distinctive soils. Volcanic ash soils or andosols possess unique properties (Shoji *et al.* 1993). They are characterized by a low bulk density (<0.85 g/cm^3) of the fine earth fraction in the epipedon or cambic horizon or both; and a resistant, dark-coloured material produced by allophane (created by the weathering of volcanic ash) combining with humus, which tends to produce the low bulk density and is the dominant source of the exchange complex. Detailed studies of volcanic soils have been made in New Zealand. Much of North Island is covered by materials erupted from volcanoes, either as tephra or as lava. These materials are rich in fresh feldspar and ferromagnesian minerals that decompose readily to provide a considerable supply of sesquioxides. It is partly this enrichment in sesquioxides that distinguishes volcanic soils from their zonal counterparts. The soils formed on North Island vary according to the age of the volcanic deposits. Two time sequences have been identified (Gibbs 1980). The first sequence is from soils formed on recently erupted volcanic material, to yellow-brown pumice soils, to yellow-brown loams, to brown granular loams. Even in the oldest of these soils, it is possible to detect the influence of the original rock material on pedogenesis. The parent materials leading to the pedosequence are shown in Figure 6.2. In many places, these materials are superimposed and buried soils are common (Plate 6.1). The youngest soils, all less than 1,200 years old, are formed in Ngauruhoe, Tarawera, Rotomahana, Burrell, and Rangitoto ashes (Table 6.1). They are all raw soils with limited horizon evolution expressed as AC profiles. The rate of pedogenesis depends on the size of volcanic ejecta, coarser particles forming soil more slowly than finer ones. It depends too on climate – the warmer the conditions the faster the rate of pedogenesis. Forming in active volcanic environments, these soils are frequently buried by deposits ejected at later dates. A thin layer of volcanic material will retard pedogenesis, whereas a thick layer will stop it and so create a buried soil. The properties of these recent soils is as variable as the profiles: some, for example, have a good supply of cations, others do not. The older yellow-brown pumice soils are better developed than the recent soils. They all have A, B, and C horizons, and some have O, E, Bh, and Bfe horizons as well. Mineral colloid content is low and chiefly consists of allophane. The level of available metal cations is low. Older still are the yellow-brown loams. These have between 10 to 20 per cent mineral colloids (mainly allophane) in each horizon, and high cation exchange capacities. The much older brown granular soils have well developed A, AB, B, BC, and C horizons (and some intergrade profiles have E and Bg horizons as well). Cation exchange capacity is moderate in the A horizon but diminishes down the profile to very low values in the bottom layer. Clay colloid content is high at 30 to 80 per cent, the main types being kaolin and halloysite.

The second time sequence of soils identified in New Zealand is brown

Plate 6.1 Three superimposed volcanic ashes near Tikitiri, New Zealand.
Photograph by Brian S. Kear

142

Table 6.1 Some volcanic soils in North Island, New Zealand

| Soil | Parent material | |
	Source and date	Texture and composition
Ngauruhoe	Mts Ngauruhoe and Ruapehu over last 400 years	Andesitic ash sands
Tarawera	Mt Tarawera, 1886	Basaltic lapilli and ash
Rotomahana	Lake Rotomahana, 1886	Rhyolitic sand, silt, and gravel
Burrell and Tahurangi	Vents on Mt Egmont between 250 and 400 years ago	Andesitic lapilli and ash
Rangitoto	Mt Rangitoto between 500 and 800 years ago	Basaltic ash
Yellow-brown pumice	Centres in Taupo, Rotorua, and Bay of Plenty between 900 and 5,000 years ago	Pumiceous rhyolite tephra, *in situ* and reworked
Yellow-brown loams	Various sources between 3,000 and 15,000 years ago	Andesitic ash, rhyolitic ash, or mixtures of the two, both as *in situ* deposits and reworked as alluvium and colluvium
Brown granular	Various sources more than 20,000 years ago	Basic igneous materials, chiefly andestic ash

loams developed on massive basaltic lava (Gibbs 1980). Three stages are recognized and represented as three soil types: Kiripaka, Ruatangata, and Okaihau. Kiripaka soils are the youngest. They are moderately leached with between 30 to 50 per cent clay, mainly consisting of kaolin and gibbsite with small amounts of micas, vermiculite, and amorphous oxides. Older soils, including the Ruatangata friable clay, are moderately to strongly acid, well leached of cations, and contain 50 to 80 per cent clay, almost all of which are kaolin and gibbsite. On ancient and stable undulating plateaux, soils including the Okaihau are moderately to strongly acid, thoroughly leached, and contain 30 to 50 per cent clay, chiefly gibbsite with a little kaolin and small nodules of iron, aluminium, and manganese oxides. These nodules are comparable to laterite in other areas and, where much alumina is present in them, they can be classed as bauxite.

Lithosequences

The influence of parent material on soils may be explored by comparing soils derived from various parent materials under comparable environmental conditions; in other words, by holding all state factors constant save parent material. This process may unveil a lithosequence, sometimes referred to as a geosequence (e.g. J. L. Richardson and Edmonds 1987), which may be expressed mathematically as a lithofunction. Lithosequences are difficult to express quantitatively because it is not easy to assign meaningful numbers to rock types. Most lithofunctions compare soils on a few rock materials. For instance, the amount of clay formed during pedogenesis may be calculated in cases where parent materials differ but other soil-forming factors can be held constant (Barshad 1958). Similarly, it is possible to compare the relative migration and residual accumulation of some chemical elements in soils in eluvial portions of landscapes developed in different parent rocks (e.g. Koinov *et al.* 1972).

Some progress has been made in singling out the pedogenetic response to a 'gradient' in a parent material. In an alluvial system, stream energy may be defined as a state variable that influences the texture of alluvial deposits. It

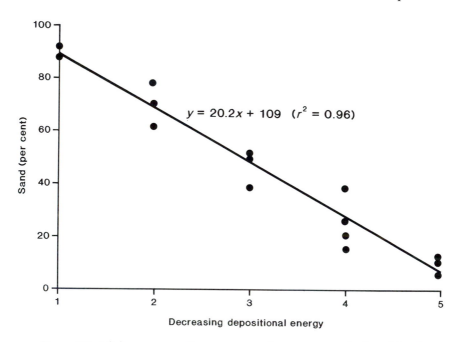

Figure 6.3 A lithosequence of percentage sand versus energy in depositional environments for the Meherrin River, Virginia. The regression equation is significant at $p = 0.05$
Source: After J. L. Richardson and Edmonds (1987)

should be possible to use depositional energy as a predictor of particle size in alluvial sediments. To test this idea, sediments in a modern floodplain were examined along the Meherrin River, Virginia, United States (J. L. Richardson and Edmonds 1987). The energy level of various sites was established by investigating primary sedimentary structures, such as cross bedding and grading. Five sites of decreasing depositional energy were identified, the full set forming a lithosequence. The sites were, in order of decreasing depositional energy: the high energy deposits on a point bar; the lower energy deposits on a point bar and the higher energy deposits on a natural levee; natural levee deposits away from the river; backswamp deposits nearest the river; and backswamp deposits farthest from the river. Percentage sand, y, was plotted against depositional environment, x (ranked from highest to lowest), and a linear regression line fitted (Figure 6.3). The relationship was

$$y = 20.2x + 109$$

The steep negative slope and the high coefficient of determination (96 per cent) indicate a significant and effective influence of depositional environment on soil texture. Another example of a lithosequence involves soils

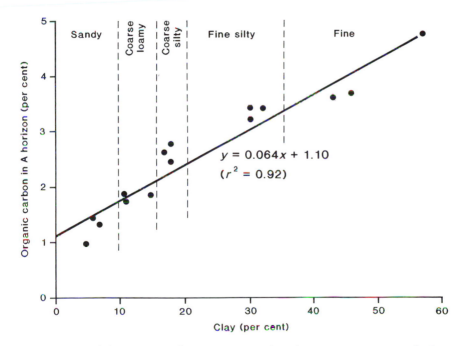

Figure 6.4 A lithosequence of percentage organic carbon versus percentage clay in the top 15 cm of soils in North Dakota (mostly Aeric Calciaquolls). The regression line is significant at $p = 0.05$. Family textural placement is noted
Source: After J. L. Richardson and Edmonds (1987)

formed in glacio-lacustrine sediments in North Dakota. Percentage clay, the independent variable (or parent-material state factor), is used to predict the organic carbon (per cent) in the top 15 cm of soil (the A horizon) (J. L. Richardson and Edmonds 1987). A relationship might be expected because clay should absorb and help to preserve organic matter from oxidation. A scattergraph and linear regression of the data confirm the relationship (Figure 6.4).

ROCKS AND LIFE

Rock undoubtedly influences animals and plants, both at the level of individual species and at the level of communities. It does so directly, where rock is exposed at the land surface, and indirectly, where the effects of rock are felt through the soil cover.

Species and substrate

Plants seem capable of adapting to the harshest of substrates. Saxicolous vegetation grows on cliffs, rocks, and screes, some species preferring rock crevices (chasmophytes), others favouring small ledges where detritus and humus have collected (chomophytes). In the Peak District of Derbyshire, England, spleenwort (*Asplenium* sp.) is a common chasmophyte and the wall-flower (*Cheiranthus cheiri*) is a common and colourful chomophyte (P. Anderson and Shimwell 1981: 142). Perhaps the most extreme adaptations to a harsh environment are seen in the mesquite trees (*Prosopis tamarugo* and *Prosopis alba*) that grow in the Pampa del Tamagural, a closed basin, or salar, in the rainless region of the Atacama Desert, Chile. These plants manage to survive on concrete-like carbonate surfaces (Ehleringer *et al.* 1992). Their leaves abscise and accumulate to depths of 45 cm. Because there is virtually no surface water, the leaves do not decompose and nitrogen is not incorporated back into the soil for recycling by plants. The thick, crystalline pan of carbonate salts prevents roots from growing into the litter. To survive, the trees have roots that fix nitrogen in moist subsurface layers, and extract moisture and nutrients from ground water at depths of 6 to 8 m or more through a tap root and a mesh of fine roots lying between 50 and 200 cm below the salt crust. A unique feature of this ecosystem is the lack of nitrogen cycling.

Vascular plants on talus at high altitudes tend to cluster around stones. This could be because the areas of rock accumulation are relatively more stable than areas of thin, fine-grained talus. Another possibility is that soil moisture is more readily available between and below stones where trapped fine-grained material holds water. Differences in temperature might also affect plant distribution. Francisco L. Pérez (1987, 1989, 1991) conducted a revealing study of soil moisture and temperature influences on the

Plate 6.2a A seedling of *Coespelitia timotensis* Cuatr., 33 cm tall, rooted between granitic fragments on a blocky talus area at an elevation 4,340 m in the Páramo de Piedras Blancas, Venezuela, January 1990. The pocket knife is 8 cm long.
Photograph by Francisco L. Pérez

Plate 6.2b A group of *Coespelitia timotensis* Cuatr. growing at the base of a blocky talus slope (with an inclination of about 31°) at an elevation of 4,360 m in the Páramo de Piedras Blancas, Venezuela, January 1990. The rocky outcrops at the talus apex can be seen in the background.
Photograph by Francisco L. Pérez

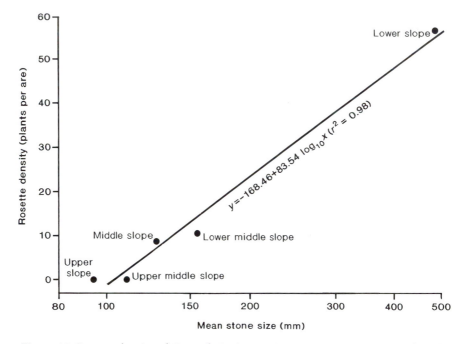

Figure 6.5 Rosette density of *Coespeletia timotensis* versus average stone size (length of longest axis in mm) on sandy and blocky talus slopes in the Páramo de Piedras Blancas, Venezuela. The regression line is significant at $p > 0.001$
Source: After Pérez (1991)

distribution of *Coespeletia timotensis*, a giant caulescent (stemmed) Andean rosette plant, on sandy and blocky talus slopes in the Páramo de Piedras Blancas, Venezuela (Plate 6.2). Rosette density and cover increased down the talus slope in parallel with increasing particle size and substrate stability (Figure 6.5). Rosette density was not so much associated with slope position *per se*, as with the increasing proportion of the talus surface occupied by large rocks downslope. The rosette plants were virtually all confined to blocky talus, and to areas downslope from isolated boulders embedded in finer sandy material. The roots of the plants always grew upslope and beneath stones. Water content of the surface soil was always ten to twenty times greater under blocky talus and beneath boulders than in contiguous areas of bare sandy talus. The amount of water available for plant growth was also higher beneath stones, even 20 cm into the soil. Soil texture was similar (sand to sandy loam) on both talus types. The extra water found under the stones could result from any or all of three processes. First, while rain is falling, water flows over the stones and accumulates in the sandy soil matrix between and under the stones. Second, after the rain has stopped falling, evaporative loss is reduced by the stone layer which prevents

148

capillary rise to the talus surface. Third, at sunset, falling temperatures promote condensation in the hollow spaces between the stones. The rosettes favour blocky talus areas because water is available under stones, even through the dry season. The bare sandy talus areas are more difficult to colonize because they become desiccated during the dry season.

Some animals are affected by landscape substrate. For instance, the type and texture of soil or substrate is critical to two kinds of mammal: those that seek diurnal refuge in burrows, and those that have modes of locomotion best suited to relatively smooth surfaces. Burrowing species, which tend to be small, may be confined to a particular kind of soil. Heteromyid rodents, for instance, are weak diggers and can live only in sandy soil. Some woodrats (*Neotoma*) build their homes exclusively in cliffs or steep rocky outcrops. The dwarf shrew (*Sorex nanus*) seems confined to rocky areas in alpine and subalpine environments. Even some saltatorial species are adapted to life on rocks: the Australian rock wallabies (*Petrogale* and *Petrodorcas*) leap adroitly among rocks and are aided in this by granular patterns on the soles of their hind feet that increase traction. Pikas (*Ochotona*) are normally resident on talus or extensive piles of gravel. The Rocky Mountain pika (*Ochotona princeps*) living in the vicinity of Bodie, a ghost town in the Sierra Nevada, utilizes tailings of abandoned gold mines (A. T. Smith 1974, 1980). The entire lifestyle of African rock hyraxes (*Heterohyrax*, *Procavia*) is built around their occupancy of rock piles and cliffs (Plate 6.3). Most of their food consists

Plate 6.3 A rock hyrax (*Procavia johnstonii mackinderi*) at an elevation of about 4,200 m on Mt Kenya.
Photograph by Nigel E. Lawson

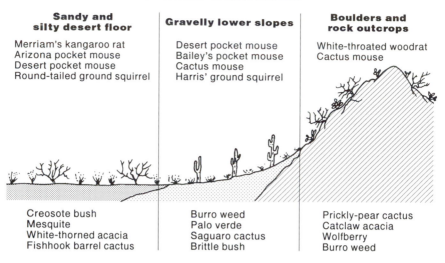

Sandy and silty desert floor	Gravelly lower slopes	Boulders and rock outcrops
Merriam's kangaroo rat Arizona pocket mouse Desert pocket mouse Round-tailed ground squirrel	Desert pocket mouse Bailey's pocket mouse Cactus mouse Harris' ground squirrel	White-throated woodrat Cactus mouse
Creosote bush Mesquite White-thorned acacia Fishhook barrel cactus	Burro weed Palo verde Saguaro cactus Brittle bush	Prickly-pear cactus Catclaw acacia Wolfberry Burro weed

Figure 6.6 Communities of mammals and plants living on contrasting substrates in
North American deserts
Source: After Vaughan (1978)

of plants growing among, or very close by, rocks. Their social system is
bonded by the scent of urine and faeces on the rocks. The rocks provide
useful vantage points to keep an eye out for predators, hiding places, and an
economical means of conserving energy. Terry A. Vaughan (1978: 431)
reports observing a colony of *Heterohyrax brucei* on the Yatta Plateau,
southern Kenya, in July 1973. On first emerging from their nocturnal
retreats, deep in rock crevices, they avoid touching the cool rock surfaces
with their undersides, turn broadside to catch the first rays of the Sun, and
bask. As the temperature soars during the morning, the hyraxes move to
dappled shade beneath the sparse foliage of trees or bushes. When the
ambient temperature tops 30 °C, they move to deep shade where they lie
sprawled on the cool rock and remain there during the hot afternoon. Before
dark, they move to the open and lie full length on warm rock.

Many desert rodents display marked preferences for certain substrates. In
most deserts, no single species of rodent is found on all substrates; and some
species occupy only one substrate. Four species of pocket mice (*Per-
ognathus*) live in Nevada (Hall 1946). Their preferences for soil types are
largely complementary: one lives on fairly firm soils of slightly sloping valley
margins; the second is restricted to slopes where stones and cobbles are
scattered and partly embedded in the ground; the third is associated with the
fine, silty soil of the bottomland; and the fourth, a substrate generalist, can
survive on a variety of soil types. Different assemblages of rodents occur
along catenae in deserts owing to variations in substrate (Vaughan 1978: 276).
A typical catena comprises rocky outcrops with boulders, gravelly lower
slopes, and a sandy desert floor (Figure 6.6). The preferences of different

150

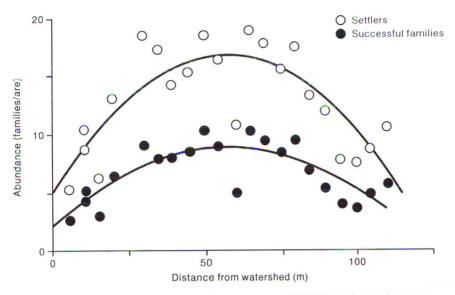

Figure 6.7 The abundance of settlers and successful families of *Hemilepistus reaumuri* on the rocky slope in the Negev Desert. Each dot and circle is a fifteen-year mean value (1973–1987). Both curves are significant at $p < 0.001$
Source: After Shachak and Brand (1991)

species for these three substrates reflect digging ability, style of locomotion, and foraging technique. Merriam's kangaroo rat (*Dipodomys merriami*) avoids danger by fast and erratic hops. Although it will live on several kinds of terrain, it prefers open ground where its mastery of hopping can be exercised to the full. In contrast, the cactus mouse (*Peromyscus eremicus*) cannot run fast, but is good at climbing and scrambling. For this reason, it lives in rocky terrain, or sometimes in areas with cactus or brush, where safe havens are within clambering distance.

Some invertebrate populations are influenced by substrate. Moshe Shachak and Sol Brand (1991) studied two populations of the desert isopod, *Hemilepistus reaumuri*, over fifteen years in the Negev Desert. The populations live in distinct landscapes: a rocky slope (spatially heterogeneous) and a loessal valley (spatially homogeneous). An aim of the study was to see how abundance and variability in the two populations related to the structure (heterogeneity versus homogeneity) of the landscape. Crucial to the survival of the isopod is soil moisture of at least 6 to 10 per cent maintained at a depth of 40 to 60 cm all summer. The study site in the rocky slope covered 11,325 m² and was divided into three sections. First, an upper rocky unit with small patches of shallow soil bearing 3 to 8 per cent vegetation cover. Second, a mid-slope unit with rock exposed over 60 to 80 per cent of the surface, soil occurring in strips at the base of bedrock steps, joints, and bedding planes, and a 5 to 10 per cent cover of vegetation, mainly along the soils strips and soil-filled joints. Third, a lower slope unit covered by an extensive stony soil, more than 250 cm thick at the slope base, and with a vegetation cover of 5 to 10 per cent. In contrast, the loessal valley site occupies 8,000 m² and is blanketed by a thick sierozem soil with a 10 to 15 per cent vegetation cover. Heterogeneity was measured by the rock–soil ratio, which was 0 for the loessal valley and ranged between 0.04 and 3.00 on the rocky slopes, and the runoff–rainfall ratio, which was a measure of the temporal heterogeneity of the water supply in the landscape.

The abundance and variability of *Hemilepistus reaumuri* were determined by measuring the densities of settlers (the number of new families formed during February) and successful families (the number of families that survived after settling until the next settling season) in two areas for fifteen generations (1973–1987). Survivorship was defined as the ratio of successful families to settlers. Variability was defined by the coefficient of variation, which is independent of the mean value. The variables were determined at the intersection points of permanent grid systems consisting of rows 5 m wide. Abundance in the loessal valley showed no trend along rows, indicating a uniform distribution of settlers and successful families. On the rocky slope, the abundance of settlers and successful families increased from upper slope to mid-slope, and then decreased towards the lower slope (Figure 6.7). No temporal pattern in the abundances at either site was detected. Variabilities for individual years among rows for settlers and successful families were significantly lower in the loessal area than on the rocky slope ($p < 0.001$ in both cases). An implication of this is that *Hemilepistus reaumuri* perceived the loessal landscape as the more homogeneous during a specific year. The isopods appear to have had greater difficulty in finding safe sites in the loessal area than they did on the rock slope. This is suggested by the insignificant difference in the variability between successful families and settlers in the rocky area ($p > 0.05$), compared with the significant difference between the relatively high variability of successful families and low variability of settlers in the loessal valley ($p < 0.001$). Temporal variability was higher in the loessal valley where the water regime is more heterogeneous.

Several conclusions were drawn from the study. First, landscape heterogeneity and population variability are positively related. Second, spatial and temporal heterogeneity are interrelated, their combined effect either increasing or decreasing variability (on the rocky slope, the high spatial heterogeneity decreases the temporal population variability, whereas the spatial location of the loessal valley increase temporal population variability). Third, the relationship between abundance and variability depends on landscape characteristics: in the loessal valley, abundance and variability are positively related, while on the rocky slope they are inversely related; and organisms respond to landscape heterogeneity by selecting sites, the stability of survivorship depending on the degree of site selection. Perhaps the most revealing conclusion of all was that relations among landscape heterogeneity, population abundance, and population variability depend on specific processes that integrate these variables: no general predictive relation could be established.

Lithobiomes

Within zonobiomes, there are areas of intrazonal and azonal soils that, in some cases, support a distinctive vegetation. Walter and Siegmar-W. Breckle (1985) have designated these non-zonal vegetation communities pedobiomes, and distinguish several on the basis of soil type: lithobiomes on stony soil, psammobiomes on sandy soil, halobiomes on salty soil, helobiomes in marshes, hydrobiomes on waterlogged soil, peinobiomes on nutrient-poor soils, and amphibiomes on soils that are flooded only part of the time (e.g. river banks and mangroves). Pedobiomes commonly form a mosaic of small areas and are found in all zonobiomes. There are instances where pedobiomes are extensive: the Sudd marshes on the White Nile which cover 150,000 km²; fluvio-glacial sandy plains; and the nutrient poor soils of the Campos Cerrados in Brazil.

A striking example of a lithobiome is found on serpentine. The rock serpentine and its relatives, the serpentinites, are deficient in aluminium. This leads to slow rates of clay formation, which explains the characteristic features of soils formed on serpentinites: they are highly erodible, shallow, and stock few nutrients. These peculiar features have an eye-catching influence on vegetation (Brooks 1987; Baker *et al.* 1992). Outcrops of serpentine support small islands of brush and bare ground in a sea of forest and grassland. These islands are populated by native floras with many endemic species (R. H. Whittaker 1954). In a locality some 5.5 km north of Geyserville, California (Jenny 1980: 248), rocky outcrops of schist support oak trees (*Quercus agrifolia*), while soils derived from schists on the extensive slopes, known as the Raynor Series, carry native bunch grass (*Stipa* sp.) and wild oats (*Avena* sp.). Adjacent soils derived from serpentine, called the Montara Series, support a forest of digger pine (*Pinus sabiniana*). The junction between schist and serpentine is sharp – no more than a metre wide. Grass in the oak-savanna grows to a height of 40 to 110 cm, then, a mere stride away, plummets to 5 to 15 cm in the digger-pine forest. The effect of serpentine on vegetation is clearly seen in Cuba. In a transect across Monte Libano, a serpentine overlies limestone (Figure 6.8). The vegetation on the limestone changes from *Roystonea–Samanea* grassland in footslopes, to semi-deciduous forest and mogote forest on backslopes. The serpentine, which underlies most of the upper slopes and summits, supports pine forest and, where bands of limestone occur, pine forest with agave. More sheltered sites on serpentine support sclerophyllous montane forest, while similar sites on limestone support submontane rain forest.

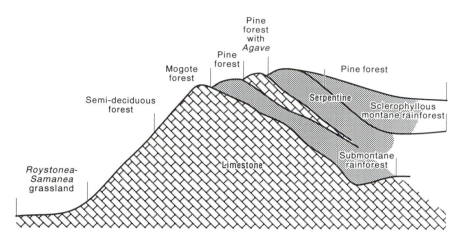

Figure 6.8 Vegetation transect across Monte Libano, Cuba
Source: After Borhidi (1991)

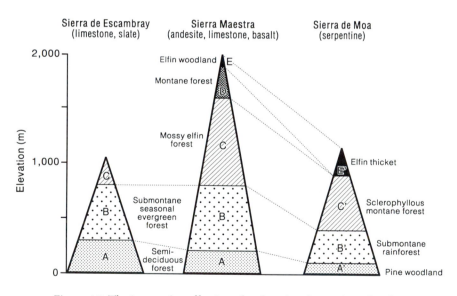

Figure 6.9 The 'serpentine effect' on the elevational zonation of Cuban vegetation. Notice that the vegetation belts on the Sierra de Moa, which are composed of serpentine, are lower than on mountains composed of other rocks
Source: After Borhidi (1991)

Alpine and montane plant species tend to occur lower down mountainsides on serpentine than they do on other rocks. In Cuba, elevational belts of vegetation are shifted wholesale down mountains in serpentine highlands (Borhidi 1991). The very humid vegetation zones, such as the cloud forest, are missing from mountain ranges formed on serpentine, and drier variants of the uppermost vegetation belts occur at half the altitude than they do on non-serpentine mountain ranges (Figure 6.9). Rock type, acting through edaphic factors, strongly influences virtually all Cuban forest and shrubwood vegetation. Relationships between vegetation and soils were established using principal component analysis of data from 267 relevés taken in 40 vegetation units producing a data matrix containing nearly 85,000 entries in more than 2,000 rows (taxa) and 40 columns (communities) (Borhidi 1991). Forty-five per cent of the data set's variance was accounted for by the first five principal components. The first component separated communities on montane serpentine in Oriente, a flora almost unique and confined to ferritic soils, from the rest. The second component separated the broadleaved forests growing mainly on limestone or neutral soils from the rest; and

Plate 6.4 An open woodland 'tree island' at 1,675 m on the southern foothills of Peavine Mountain, Nevada. The conifers growing on the light-coloured, hydrothermally altered andesite are *Pinus ponderosa*, *P. jeffreyi*, and *P. lambertiana*. The surrounding vegetation, which grows on brown desert soils, is typical sagebrush (*Artemisia tridentata*) with associated species, *Purshia tridentata* and *Ephedra viridis*.
Photograph by William H. Schlesinger

155

the third principal component had high positive loadings for lowland shrubwoods and pinewoods growing on serpentine. Ordination of the data pointed to the overriding influence of edaphic factors in explaining variations in Cuban forest and shrubwood vegetation, although rain forests that are influenced chiefly by climate were unaffected by the edaphic factors.

A remarkable lithobiome is found in the western Great Basin desert of the United States where 'tree islands' grow in a sea of sagebrush vegetation (Plate 6.4). The islands take the form of about 140 small stands of Sierra Nevada conifers (mainly *Pinus ponderosa* and *Pinus jeffreyi*), from one to several hectares in area, lying up to 60 km east of the eastern margins of the Sierra Nevada montane forest. They are restricted to outcrops of hydrothermally altered andesitic bedrock, from which base cations have been leached on exposure, that produce localized patches of azonal soil (Billings 1950). The soils derived from andesitic bedrock in the Great Basin are primarily Xerollic Haplargids, typical of desert brown soils, whereas the soils derived from the altered bedrock form shallow Lithic Entisols, light-yellow in colour, acid in reaction, and low in base cations and phosphorus (DeLucia and Schlesinger

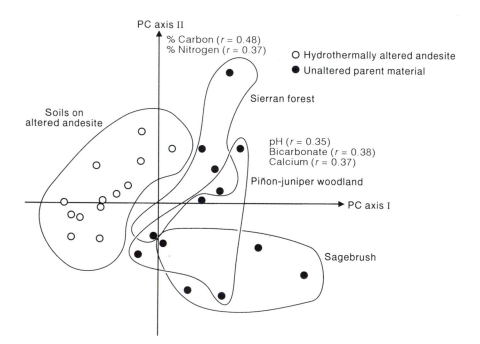

Figure 6.10 Principal component analysis of soil data from vegetation in and around 'tree islands' of the western Great Basin Desert, United States. The first two axes account for 48 per cent of the variance and are most strongly correlated with the variables indicated
Source: After DeLucia and Schlesinger (1990)

1990). Principal component analysis of eighteen soil parameters indicated that soils formed in altered rock from different sites have much in common, but soils formed in unaltered rock differ according to vegetation cover – forest, piñon–juniper woodland, or sagebrush (Figure 6.10) (DeLucia *et al.* 1989; Schlesinger *et al.* 1989). The first principal component relates to acidity, alkalinity, and calcium content; the second to carbon and nitrogen content. Taken together, these two axes account for 48 per cent of the variance in the data.

Lithobiomes are associated with the talus slopes common in alpine, Arctic, and desert regions. Talus is formed by the accumulation of loose rock debris of varying sizes. Plants appear to have difficulty in colonizing talus. Where colonization has taken place, plants are commonly associated with specific talus zones or substrate types. In the Jura Mountains, central Europe, Roman Bach (1950) found that talus slopes formed of limestone fragments are graded: the small fragments accumulate beneath rock outcrops, the source of the talus, while the biggest (blocks with diameters of about 50 cm) lie at the foot of the talus slope. This gradation of particle size creates a lithosequence of parent materials, soils, and vegetation. On the upper slope, rendzina soils evolve. They are a deep, gravel-rich sandy loam with a granular structure topped by 60 to 100 cm of mull humus. Their pH ranges from 6.5 to 7.8. These productive soils support a forest of mountain maple (*Acer* sp.), with shrubs of ash (*Sorbus* spp.) and hazel (*Corylus* sp.), and a herb layer predominated by ferns and members of the Cruciferae. Towards the foot of the talus slope, blocky raw carbonate soils evolve. There is no fine soil material, and a layer of mor humus, some 30 cm thick, lies directly on the limestone boulders. Some organic matter is washed between the boulders and feeds roots. Spruce (*Picea abies*) forest and *Hylocomium* mosses grow in this geomorphologically active landscape. The spruce does not ascend too far up the talus slope because it cannot endure the frequent salvos of rolling boulders and the motion of the soil.

SUMMARY

Substrate exerts a strong influence on the structure and function of many geoecosystems. Parent material exerts a dominating influence on soil evolution at microscales and mesoscales. Several rocks – such as limestone, volcanic ash, serpentine – produce very distinctive soils. Soil lithosequences have been discovered, though 'gradients' within lithological 'fields' are often so extremely steep as to be discontinuities. Substrate influences individual animals and plants, as well as communities. Plants have adapted to the harshest of substrates, including the rainless region of the Atacama Desert, Chile. Animals, too, have adapted to harsh substrates. Some species, such as pikas and rock wallabies, are geared to life on the rocks. At the community level, certain vegetation types are peculiar to particular substrates. These pedobiomes include lithobiomes (vegetation on stony soil), of which a fine example is the plants associated with serpentinites. Other lithobiomes are the 'tree islands' formed on hydrothermally altered andesite in the western Great Basin, United States, and the communities associated with talus slopes.

FURTHER READING

A clear summary of substrate and its effect on soil (and to a lesser extent vegetation) is the chapter on parent material in Jenny's *The Soil Resource: Origin and Behavior* (1980). Robert Richard Brooks's tome, *Serpentine and its Vegetation: A Multidisciplinary Approach* (1987), is an excellent case study of the effects of serpentine on soils and plants. The relationships between vegetation and substrate are discussed in many of Monica Mary Cole's papers, most of which are listed in her book on *The Savannas: Biogeography and Geobotany* (1986). Information on the relations between animals and substrate is scattered in ecological and zoological journals. The references mentioned in the chapter could be used as a starting point for a literature search.

7

TOPOGRAPHY

Topography is perhaps the most conspicuous component of a geoecosystem. It may be characterized by several measures including relief, aspect, slope gradient, slope curvature, slope length, and contour curvature. The influence of relief, which exerts itself chiefly through climate, was tackled in the fifth chapter. In this chapter, the influence of aspect and slopes on organisms and soils will be investigated.

ASPECT

Aspect strongly affects the climate just above the ground and within the upper layers of the regolith. For this reason, virtually all landscapes display significant differences in soil and vegetation on adjacent north-facing (distal) and south-facing (proximal) slopes, and on the windward and leeward sides of mountains and hills. In the Northern Hemisphere, south-facing slopes tend to be warmer, and so more prone to drought, than north-facing slopes. The difference may be greater than imagined. In a Derbyshire dale, the summer mean temperature was 3 °C higher on a south-facing slope than on a north-facing slope (Rorison *et al.* 1986), a difference equivalent to a latitudinal shift of hundreds of kilometres! These differences affect pedogenesis and plant growth. They also affect the microclimate in which animals live and so influence animal distribution. Differences between the climate of windward and leeward slopes may also be consequential: large mountain ranges cast a rain shadow in their lee that is sufficient greatly to alter the vegetation and soils.

Soils and aspect

Aspect affects the evolution of soils. A host of examples attests to this fact (see Carter and Ciolkosz 1991). In the southern Appalachian mountains, red podzolic soils are found at higher altitudes on southern slopes; in the western United States, podzolic soils occur at lower altitudes, and are more common, on northern slopes (Lutz and Chandler 1946). In virgin soils in western Iowa,

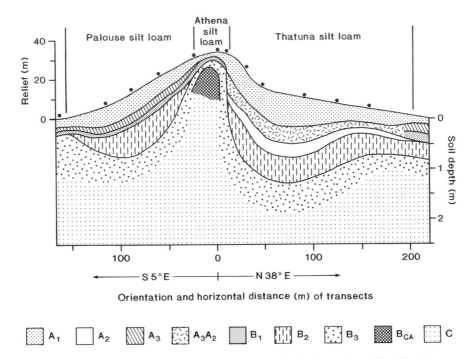

Figure 7.1 Cross section of Palouse Hill showing the horizons for soil of the Palouse Catena. The dots indicate sampling sites
Source: After Lotspeich and Smith (1953)

nitrogen content is higher in lower slope positions and on north-facing slopes, the higher contents on north-facing slopes possibly resulting from decreased evaporation there (Aandahl 1948). In Washington, the evolution of soil catenae formed in loess are influenced by variations in microclimate produced by aspect (Lotspeich and Smith 1953). The effective moisture on north-facing slopes, summits, and south-facing slopes varies considerably. South-facing slopes receive close to the annual precipitation of 53 cm, but 20 per cent of this falls as snow and is easily redistributed in the landscape. Summits may have an effective precipitation of a mere 25 cm, the remaining water being lost by higher evaporation on the exposed ridges and in snow blown by the wind. North-facing slopes have a higher effective precipitation than south-facing slopes, mainly because snow drifts accumulate there and evaporation rates are low. The result is that prairie soils (brunizems) evolve on gentle, south-facing slopes, chernozems on dry ridges, and claypan soils on moist lee slopes (Figure 7.1). In the Tanana watershed region of Alaska, south-facing slopes are mantled by a subarctic brown forest soil supporting mature white spruce; north-facing slopes are covered by a half-bog soil in which grows non-marketable black spruce (Krause *et al.* 1959). A study

160

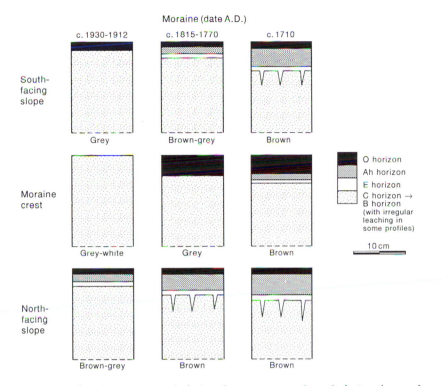

Figure 7.2 Soil evolution on north-facing slopes, crests, and south-facing slopes of moraines of different ages, Bödalsbreen, southern Norway
Source: After Matthews (1992), after a doctoral dissertation by O. R. Vetaas

carried out in the E. S. George Reserve, south-eastern Michigan, made during the growing season of 1957 revealed the influence of aspect on soil properties. Compared with soils on north-facing slopes, soils on south-facing slopes had lighter brown A horizons and redder B horizons, contained more clay in the B horizon (on average, almost 5 per cent more), had thinner A horizons, and were shallower (A. W. Cooper 1960). Ole R. Vetaas (cited in Matthews 1992) measured several soil properties on north-facing (distal) slopes, crests, and south-facing (proximal) slopes on moraines of different ages in front of Bödalsbreen, southern Norway. The data indicate that the genesis of podzols is slowest on moraine crests, and more rapid on distal slopes than on proximal slopes (Figure 7.2). On the youngest moraine, traces of a leached (E) horizon are found on the distal slope, but not on the proximal slope. On the next youngest moraines, an E horizon is present on proximal slopes as well as on distal slopes. After about 230 years, pedogenesis is more pronounced on the distal slope which has thicker Ah and E horizons. Vetaas attributes these differences on north- and south-facing slopes to the effect of

161

a strong glacier wind that reduces the accumulation of litter and the production of humus on moraine crests, and to a lesser degree on proximal slopes. The distal slopes are more sheltered from the effects of the wind and pedogenesis can work faster.

Animals, plants, and aspect

Life-forms of plants appear to be influenced by aspect, in some cases dramatically so (Plate 7.1). John E. Cantlon's (1953) study on north-facing slopes, south-facing slopes, and the entire ridge of Cushetunk Mountain, New Jersey, shows significant differences (Table 7.1). In south-eastern Ohio, microclimatic differences between north-east-facing and south-west-facing valley sides lead to a mixed-oak association on south-west-facing slopes and a mixed mesophytic plant association on the moister north-east-facing slopes (Finney *et al.* 1962). In the Front Range, Colorado, the vegetation changes from ponderosa pine in the lower mountain zone, through mixed Douglas fir and ponderosa pine in the upper montane zone, and Englemann spruce–subalpine fir forest in the subalpine zone, to *Kobresia* meadow tundra in the alpine zone (Billings 1990). Aspect influences the distribution of the trees in all zones. The ponderosa pine stands are open and park-like on south-facing slopes, but are dense on north-facing slopes and are often admixed with Douglas fir. In the zone where ponderosa pine and Douglas fir occur together, the pines are dominant on the south-facing slopes and Douglas fir on the north-facing slopes, but Douglas fir becomes more common on south-facing slopes as elevation increases. At around 2,600 m, lodgepole pine and aspen occur in fairly pure stands on north-facing slopes where fire has occurred in the past. From about 3,050 m, Englemann spruce and subalpine fir (*Abies lasiocarpa*) are dominant.

Aspect determines exposure to prevailing winds. Leeward slopes, especially on large hills and mountains, normally lie within a rain shadow. Rain-shadow effects on vegetation are pronounced in the Basin and Range

Table 7.1 The proportion of plant life-forms, based on the Raunkiaer scheme, on Cushetunk Mountain, New Jersey

Aspect	Number of species	Phanero-phytes	Chamae-phytes	Hemicryp-tophytes	Crypto-phytes	Thero-phytes
Entire ridge[a]	229	31.0	1.3	48.5	13.1	6.1
North-facing	86	41.8	1.2	41.8	15.2	0
South-facing	112	32.1	1.8	46.4	12.5	7.1

Note: [a]Total observed flora
Source: After Cantlon (1953)

Plate 7.1 The effect of aspect on plant growth, Findelen, near Zermatt, Switzerland. The forested north-facing slopes (*ubac*) contrast starkly with the alpine meadow on the south-facing slopes (*adret*).
Photograph by David N. Collins

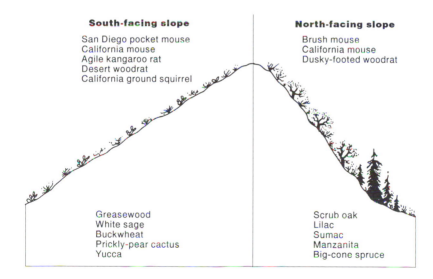

Figure 7.3 Mammal and plant communities on south-facing and north-facing slopes in lower San Antonio Canyon, San Gabriel Mountains, California
Source: After Vaughan (1978)

163

Province of the United States: the climate of the Great Basin and mountains are influenced by the Sierra Nevada, and the climates of the prairies and plains are semiarid owing to the presence of the Rocky Mountains. In the Cascades, the eastern, leeward slopes are drier than the western, windward slopes. Consequently, the vegetation changes from hemlocks (*Tsuga heterophylla* and *Tsuga mertensiana*) and firs (*Abies amabilis* and *Abies lasiocarpa*) to western larch (*Larix occidentalis*) and ponderosa pine, and finally to sagebrush desert (Billings 1990).

Geographical ranges of some animal species are influenced by aspect. In the mountainous regions of the western United States, the valleys tend to lie on an east–west axis. Consequently, the south-facing slopes are drier and warmer than adjacent north-facing slopes. These microclimatic differences strongly influence the distribution of animals and plants. For instance, in the steep-sided mountains of southern California, where the 'climax' vegetation is chaparral, the biotas on adjacent north-facing and south-facing slopes are altogether different. Some species of small mammal, such as the San Diego pocket mouse, are confined to south-facing slopes, and others, such as the dusky-footed woodrat, are restricted to north-facing slopes (Vaughan 1954) (Figure 7.3).

TOPOSEQUENCES

The catena concept

Dokuchaev noticed that the depth of chernozem profiles varies on undulating terrain. This led him to suggest that material is redistributed by topography (Joffe 1949). In 1935, Geoffrey Milne put forward the concept of the catena as a unified framework within which to study functional aspects of soils on hilly terrain. Milne was based at the East African Agricultural Research Station, Amani. A problem he faced was to map on a small piece of paper, the complex entanglement of soils in a large piece of country. To surmount the problem, he took advantage of the fact that in many parts of East Africa, the topography over large tracts consists of little else but a repetition of crests and hollows. A transect running from crest to hollow traverses very different soil profiles. Mapping the individual soils along the transect would be impossible in all but the most detailed of surveys. Nor, he reasoned, would it be necessary to do so because 'they are not, properly speaking, individual soils at all, but are a compound soil unit of another kind in which a chain of profile-differences occurs in a regular manner' (Milne 1935a: 192). Now this idea seems eminently logical today. At the time, however, it was mainly the well-drained soils of a region that were singled out as characteristic zonal types and represented on maps. The ill-drained soils in valley bottoms were demoted to intrazonal status and suppressed. Milne bravely contended that the soils of the bottomlands are as important as the soils on the ridges. The soils occurring along a crest to valley transect vary continuously, but usually three or four distinctive types can be picked out. To describe the regular repetition of soil profiles on crest–hollow topography, which forms a convenient mapping unit, Milne chose the word 'catena'. This is Latin for a chain. He considered adopting the word 'suite', as used by Gilbert Wooding Robinson to describe a range of differing soils related by topography in Wales. Robinson confined his suites to soils formed in the same parent material; he did not deal with such extreme differences in soils on hills and in valleys as Milne did. Briefly, Milne proposed the term catena to describe the lateral variation of soils on a hillslope, and reasoned that, owing to the agency of geomorphological and pedological processes, all soils occurring along a hillslope are related to one another. He was quite explicit that the topographic relationships of the soils were the prime concern, and that the uniformity of parent material was of subsidiary interest. Milne (1935a, 1935b) gave several examples of catenae from East Africa. In what was formerly central Tanganyika, now Tanzania, he described catenae formed in a series of troughs or basins having level floors, evidently old lake basins. The centre of each basin carries dark-coloured clays rich in calcium. In successive zones moving outwards towards the margins, are grey sandy clays and slightly clayey sands. The banks of the former lakes are represented by a greater or lesser development of red earths. A point of great significance made by Milne, in what was possibly the first recognition of a vegetation catena, was that each soil zone supports a characteristic type of vegetation – open grass-steppe in the middle, through various formations of grass and acacia thorn, to a mixed deciduous 'Urbusch' and baobabs on the red earths.

The catena concept had a mixed reception, but was generally hailed a valuable idea. It prompted a spate of studies considering the relationships between soils and hillslopes. The concept was embellished by Thomas M. Bushnell (1942, 1945), who also identified precedents to it, though the term 'catena' was assuredly first adopted by Milne. A potent development came from Cecil G. T. Morison (1949), then at the Department of Agriculture, University of Oxford. Morison went on several expeditions to the Anglo-Egyptian Sudan to investigate the soil–vegetation units. Preliminary work suggested that the catena concept could usefully be adopted as a framework of study in this area. He found it helpful to distinguish three zones (he termed them complexes) along a catena, each associated with a broad topographic site: the eluvial zone, the colluvial zone, and the illuvial zone. The eluvial zone is a high-level site that loses water and soluble and suspended matter. Material washed from it is used to build up the colluvial and illuvial zones. The colluvial zone occupies slope sites. It receives material from soils in the eluvial zone and loses some of it to the illuvial zone. The illuvial zone occupies low-level sites. In many cases it has very mixed parentage, consisting of either a simple mosaic or else a mosaic of zoned patterns, depending upon the amount and nature of drainage. It has three distinguishing characteristics: it receives more water than the climatic normal site; it receives much dissolved and suspended matter; and water is lost from it by surface movement, by drainage, or by evaporation.

It would seem that Morison designated his slope zones in ignorance of Boris B. Polynov's work. Polynov believed in the integrity of the landscape in producing, transporting, and removing rock debris. Two ideas are central to his thesis: first, that there are three basic landscape types relevant to chemical migration; and second, that each chemical has a characteristic mobility in the landscape (Polynov 1935, 1937). His basic landscape types were eluvial, superaqual, and aqual. In eluvial landscapes, the water table is always, or nearly always, below the ground surface; in superaqual landscapes, the water table and the ground surface coincide; in aqual landscapes, free water rests on the land surface as in lakes. From Polynov's pioneering studies have evolved several conceptual schemes of geochemical landscapes. Notable contributions have come from Mariya Al'fredovna Glazovskaya (1963, 1968). Her scheme for classifying and mapping geochemical landscape allows for the possibility that many landscapes are polygenetic. Homogeneous landscapes are formed within a single weathering cycle; heterogeneous landscapes are the production of two or more weathering cycles. As for the landscapes themselves, she adopts Polynov's basic landscape types and subdivides them (Table 7.2 and Figure 7.4). Eluvial landscapes are divided into four kinds: truly eluvial landscapes, as found on many summits; transeluvial landscapes on the upper parts of valley-side slopes; eluvial accumulative landscapes at the base of valley-side slopes; and accumulative eluvial landscapes in valley bottoms where the layer of accumulated material is deep. Superaqual and aqual landscapes she divided into two groups – those with running water and those with stagnant water.

Topography was one of Hans Jenny's factors of soil formation. Jenny argued that a catena runs from crest to crest across an intervening valley. The sequence from crest to valley bottom, what could be called a half catena, he named a toposequence (short for topographic sequence). Milne used the word 'catena' for half catena, and most people use it in that way. Technically speaking, a catenary curve, such as would be produced by suspending a chain between two points, does hang from a high point, through a low point, to a high point. None the less, there seems no reason why the words 'catena' and 'toposequence' cannot be used interchangeably.

Most environments display some signs of geomorphological activity that will influence biological and pedological processes. Soils and vegetation seldom develop in a totally inactive geomorphological environment: the landscapes in which soils and vegetation develop normally change. Thus the development of terrestrial ecosystems and the geomorphological development of landscapes take place at the same time and influence one another. Pedologists were alerted to this fact by Shaw (1930), who openly included

Table 7.2 Glazovskaya's classification of landscape elements

Polynov's basic landscape types	Landscape type according to geochemical processes	Landscape type according to stage in rock cycle		
		Primary	Secondary	Superimposed secondary
Eluvial	Eluvial (summits and well-drained, ancient plains)	Ortheluvial (on massive igneous rocks)	Paraeluvial (on dense sedimentary rocks)	Neoeluvial (on unconsolidated sediments)
	Transeluvial (shoulders and backslopes)	Transortheluvial	Transparaeluvial	Transneoeluvial
	Eluvial accumulative (footslopes and dry toeslopes)	Transortheluvial accumulative	Transparaeluvial accumulative	Neoeluvial accumulative
	Accumulative eluvial (confined footslopes and toeslopes with a low water table)	Orthoaccumulative eluvial	Para-accumulative eluvial	Neoaccumulative accumulative
Superaqual	Trans-superaqual (transhydromorphic)	Transorthohydromorphic	Paratranshydromorphic	Neotranshydromorphic
	Superaqual (confined lower ground with limited exchange of water)	Orthosuperaqual	Parasuperaqual	Neosuperaqual
Subaqual	Transaqual (streams and flowing lakes)	Transaqual	Transaqual	Transaqual
	Aqual (stagnant lakes)	Aqual	Aqual	Aqual

Source: After Glazovskaya (1963)

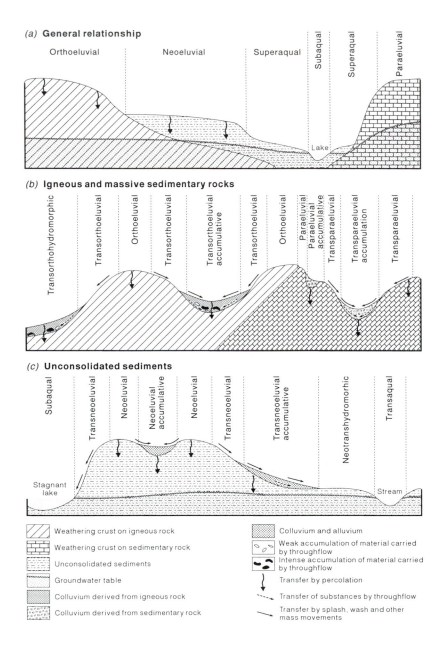

Figure 7.4 Glazovskaya's geochemical landscape elements. (a) General relationships. (b) Igneous and massive sedimentary rocks. (c) Unconsolidated sediments

Source: After Glazovskaya (1963)

erosion and deposition as 'factors' of soil formation, and by Robinson (1936) and Milne (1936), who discussed the role of 'normal' erosion in soil evolution. Until recently, few pure pedologists considered the significance of Earth surface processes to pedogenesis; that task was, with some notable exceptions (e.g. Conacher and Dalrymple 1977; R. B. Daniels and Hammer 1992), left to geomorphologists with an interest in soils. Landscapes are dynamic systems and appear to have changed appreciably, even during the Holocene epoch (Gerrard 1991), and so will have influenced pedogenesis. Interestingly, Adrian E. Scheidegger (1986) has enunciated a catena principle in geomorphology which holds that all landscapes may be viewed as a collection of catenae, and that each catena comprises an eluvial, colluvial, and alluvial zone. The eluvial zone lies at the top of the catena and consists of a flat summit and shoulder; the colluvial zone lies in the middle of the catena and consists of a backslope and footslope; the alluvial zone lies at the bottom of the catena – it is the toeslope which, like the summit, is fairly flat. This wider application of the catena concept at least serves to draw attention to the unitary nature of geoecosystems.

Lateral movement of soil materials

During the 1960s, researchers started to take up Milne's and Morison's seminal ideas and investigate soil evolution in the context of landscapes. Several statements affirmed Morison's view: soils on lower slopes are potential sumps for the drainage of soils upslope of them (Hallsworth 1965); on hilly terrain, water movement connects soils with one another and differentiates their properties (Blume 1968); adjacent soils at different elevations are linked by a lateral migration of chemical elements to form a single geochemical landscape (Glazovskaya 1968). A cardinal point is that, because solution and water transport act selectively, the lateral concatenation of soils leads to the differentiation of soil materials. This means that the hill soils in a landscape may be thought of as A horizons, and the valley soils as B horizons (Blume and Schlichting 1965). These ideas, fostered by a consideration of the catena concept, have led to the lateral translocation of soil material along toposequences being investigated.

Lateral translocation may be assessed in the field by sampling and analysing mobile soil materials. A study in south-eastern Saskatchewan revealed that salts derived from summits had been carried downslope during periods of abnormally high precipitation and had accumulated in the toeslope soils (Ballantyne 1963). In Hettinger County, North Dakota, water in excess of crop use leaches soluble salts from the root zone and transports them by overland flow and throughflow to lower landscape positions where they accumulate as saline seeps. Evaporation of water from the seeps causes the dissolved salts to rise through the soil, resulting in salt-crust formation at the surface (Timpson *et al.* 1986). Near Cape Thompson, Alaska, very

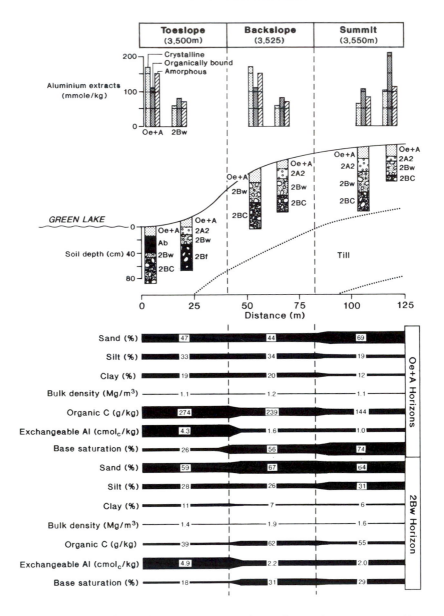

Figure 7.5 A geochemical catena at Green Lakes Valley study site, near Boulder, Colorado
Source: After Litaor (1992)

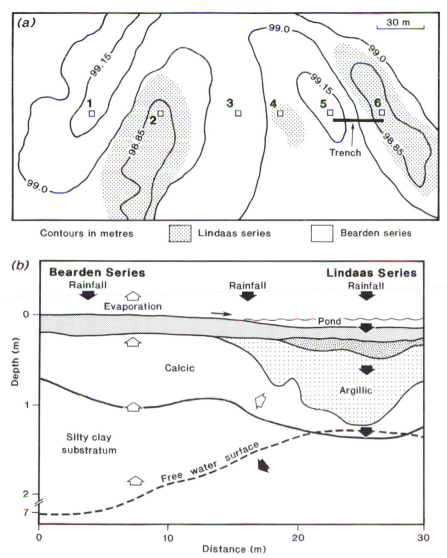

Figure 7.6 A study site within the Bearden–Lindaas soil complex in Traill County, North Dakota. (a) Topography and soil moisture and temperature stations (marked by squares). (b) A diagrammatic model of the flow of water in the soils along the trench marked in (a)

Source: After J. L. Richardson *et al.* (1992) and Knuteson *et al.* (1989)

poorly drained soils occupying low ground have a higher burden of strontium-90 than better drained soils on high ground, probably because the strontium-90 has been washed downhill (Holowaychuk *et al.* 1969).

More recent work on soluble material in soil catenae tends to link translocation of soil material to detailed studies of hillslope hydrology (e.g. Conacher 1975; Muhs 1982; Durgin 1984; Hauhs 1986; Knuteson *et al.* 1989; Arndt and Richardson 1989, 1993; Hopkins *et al.* 1991; J. L. Richardson *et al.* 1992). A good example of this is M. Iggy Litaor's (1992) investigation of lateral aluminium movement in an alpine drainage basin. A catena was studied within Green Lakes Valley, 35 km west of Boulder, Colorado. Soils and soil solutions were studied to assess the mobility of aluminium along the catena. It was found that organic carbon, exchangeable aluminium, silt, and clay increase downslope, whereas base saturation and soil buffering capacity decrease (Figure 7.5). This physiochemical pattern appears to result from enhanced lateral flow within surface horizons over frozen subsurface horizons during the snowmelt season. Total reactive aluminium, total monomeric aluminium, and hydrogen ions became more concentrated in subsurface horizons after a major summer-storm event, probably owing to vertical leaching coupled with throughflow. Throughflow may translocate clay along the catena, a process noticed earlier in a small drainage basin in the Northaw Great Wood, England (Huggett 1976). On a smaller scale, the genesis of calcic horizons on the Lake Agassiz Plain, eastern North Dakota, has been explained by groundwater recharge influenced by microtopography and a silt-clay substratum (Knuteson *et al.* 1989). Two soil series occur in the study site: Bearden soils have a near-surface calcic horizon, while Lindaas soils have a thick argillic horizon underlain by a carbonate-bearing horizon below 1 m (Figure 7.6). Detailed topographic and soils maps were prepared, and a 210-m transect with six soil moisture and temperature stations was laid out in the Bearden–Lindaas mapping unit (Figure 7.6a). After sixty weeks of monitoring, a 33-m long linear trench was excavated parallel to the transect for morphological investigations and sampling. Analysis of the data suggested that water flows through the landscape system in the manner depicted on Figure 7.6b. Surface water flows to the depression, causing ponding in the spring and early summer, then recharges groundwater, which moves laterally and upwards in response to hydraulic gradients. The present concentration of calcium carbonate equivalent, 3.5 mmol/kg, in the rising groundwater lies within the annual range necessary to account for the observed calcium carbonate accumulation in the calcic horizon of the Bearden soils. Subsequent work using flownet analysis, which models the kinetics of water transfer, lent support to this explanation of calcic horizon formation (J. L. Richardson *et al.* 1992).

Radioactive tracers and mobile salts do not normally reveal long-term changes in catenae. Geochemists have used heavy metal concentration patterns in landscapes to trace veins of ore, and their investigations plainly show that soil material moves downslope (e.g. Rose *et al.* 1979; M. F. Thomas *et al.* 1985). However, the geochemical work appears to have proceeded independently of pedological work. In pedology, changes in soil properties during soil catena evolution may be assessed by using a reconstruction technique that quantifies gains and losses of soil constituents relative to their concentrations in the parent material (e.g. Evans 1978). An early attempt to quantify the downslope movement of soil material was made in a catena in Cass County, Illinois (Smeck and Runge 1971). In some soils of the catena, more phosphorus had accumulated in the B horizons than could be accounted for by eluviation from the superjacent A horizons, and in other soils more phosphorus had been lost from the A horizons than had accumulated in the B horizons. Net gains and losses in each profile were used to investigate phosphorus dynamics. Using zirconium oxide as an index against which to gauge changes in phosphorus content, and estimating the area represented in the landscape by each profile, absolute gains and losses for each soils unit were calculated. Summit soils had lost 151.34 g/m^2 of phosphorus; footslope soils had gained 493.49 g/m^2 of phosphorus.

Some reconstructions of catenary development suggest that lateral movement of soil materials has not been important in pedogenesis, though the profiles none the less vary according to their topographic setting. In Texas, reconstruction techniques were used to probe catenary influences on the evolution of horizons rich in carbonate (West *et al.* 1988). The conclusion was that gains and losses of carbonates in summit soils were best explained by deeper leaching in this stable landscape position. Backslope soils were unstable, and carbonate distribution with depth was affected more by erosion than by downslope enrichment with carbonates. Soil reconstruction techniques were employed to examine relationships between landscape position and soil genesis in the Piedmont and Blue Ridge Highlands region in Virginia (Stolt *et al.* 1993a, 1993b). Four toposequences were selected for detailed study. Two were on mica gneiss, one on augen gneiss, and the fourth on gneissic schist. The chief processes that could be quantified using reconstruction analysis were sand and silt weathering, subsequent transport and leaching of weathering products, clay illuviation, and the accumulation of free iron oxides. Results showed that summit and backslope soils undergo the same process of soil evolution. Footslope soil genesis is, in part, dependent on the type and composition of parent material: horizons formed in substantially weathered local alluvium displayed minimal sand and silt weathering and minimal leaching; subjacent horizons evolved in parent rock yielded signs of weathering and clay eluviation. There was evidence that some material in footslope soils was derived from upslope. However, differences between the soils on summits and backslopes, which are morphologically very similar, appeared to have resulted from soil disturbance by tree-throw or natural hillslope erosion, from accelerated erosion due to cultural practices, or differences in parent materials, and not simply from catenary position.

Catenae, as a part of geoecosystems, involve geomorphological processes as well as pedological processes (cf. Scheidegger 1986). Transfer of the clastic debris of weathering is greatly influenced by landform and position. Richard

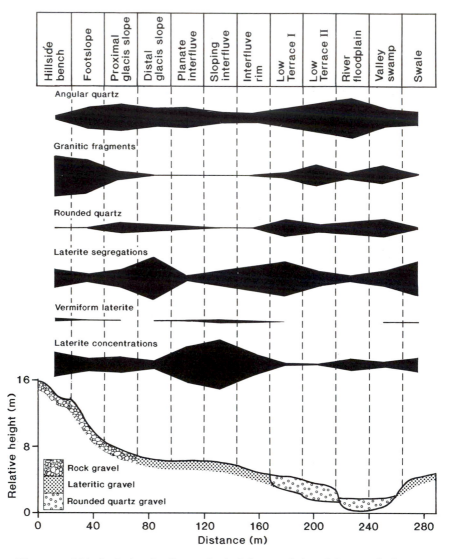

Figure 7.7 Lithological and sedimentological characteristics of the gravels along a catena in Sierra Leone
Source: After Teeuw (1989)

M. Teeuw (1989, 1991) prospected the relationship between the morphology of the land surface and the debrisphere in the forest–savanna zone of Sierra Leone. Ten transects from interfluve to valley floor were sampled by digging pits to bedrock. All landform elements along the transects had a gravel layer with characteristic average depth, texture, and petrographic composition (Figure 7.7). Examining variations in the gravel layer in a catenary framework

174

revealed the modulation of contemporary landscape processes by landscape form. Three process domains appear to interact along a catena: a residual domain, with bedrock disintegration on the hilltops and pedogenesis with biogeochemical weathering on the planate interfluves; a colluvial regime, with micro-pedimentation of ferricrete layers, surface wash (inter-rill) erosion and deposition of soil, transfer of weathered products by lateral eluviation, and precipitation of sesquioxides at breaks of slope; and a fluvial regime, with clast attrition during channelled flow, dissolution of sesquioxide compounds, and removal of fine weathering products from the drainage basin.

Other studies have tried to simulate the movement of soil materials along a catena using mathematical models. One such study (Huggett 1973) considered salts carried by throughflow in a homogeneous soil. The salts moved downslope, which is hardly surprising, but several subtle effects were generated by the simulations. Downslope movement led initially to a salt build-up at the midslope, concave–convex junction. One might expect this to be the case, but it is interesting that it has been observed in some slope soils (Furley 1971; Whitfield and Furley 1971). This 'peak' of concentration is not a static feature – as time progresses it moves as a wave down the slope. The notion of a transient concentration wave moving down a catena is not unreasonable since ions do progress through a vertical column of soils in this manner (Yaalon 1965), and may well move down a catena in the same way (Yaalon *et al.* 1974). Field evidence of the wave-like progress of peak concentration along a catena is meagre. If mobile soil materials do move down a catena in the same way as they move through a soil profile, then, as each material has a characteristic mobility, the peak of the concentration wave will appear at different positions for different materials. Concentration 'peaks' along a catena in the Northaw Great Wood, south Hertfordshire, England did show signs of this phenomenon in horizons formed in London Clay: starting upslope, the order of the peak concentrations was aluminium, iron, and manganese (Huggett 1976).

Soils and slopes

Another line of research prompted by the catena concept is the empirical investigation of relationships between soils properties and morphological variables describing hillslope profiles. Soil properties have been found to relate to slope gradient, slope curvature, and slope position, though exceptions have also been discovered (e.g. Gerrard 1988; 1992: 56–58). An interesting piece of research was carried out on the Berkshire and Wiltshire chalk downs, England (K. E. Anderson and Furley 1975). Relationships were sought between selected surface soil properties and topographic measures (slope gradient and slope length) using principal component analysis. A consistent pattern in the distribution of soil properties over five slope

Plate 7.2a Soil catena on Pinedale 2 moraine.
Photograph by David K. Swanson

Plate 7.2b Soil catena on Bull Lake 2 moraine.
Photograph by David K. Swanson

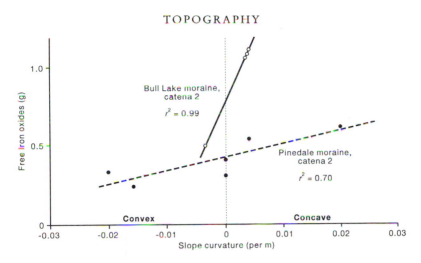

Figure 7.8 Free iron oxides (in a $1 \times 1 \times 130$ cm soil column) versus hillslope curvature ($\partial^2 x/\partial y^2$) for catenae Pinedale 2 and Bull Lake 2, Wind River Mountains, Wyoming
Source: After Swanson (1985)

transects was found. The first component of the pattern, which accounted for between 50 and 60 per cent of the total variance in soil properties, was related to organic matter and soluble constituents. Properties associated with organic matter (carbon content, nitrogen content, exchangeable potassium, and moisture loss) diminished fairly evenly downslope, whereas properties associated with soluble constituents (pH, carbonates, exchangeable calcium, sodium, and magnesium) increased downslope. The second component, accounting for 13 to 18 per cent of the variance, was interpreted as a 'particle size' factor. Values for this component showed an abrupt increase in finer soil material immediately downslope of the maximum slope gradient in the transect, giving a marked discontinuity in the pattern over the slope. In an earlier paper (Furley 1968), it had been suggested that some slopes could be divided into two sections: an upper, generally convex section where net erosion is greater than net deposition; and a lower, generally concave section where net deposition dominates. The zone of interaction between the two sections is known as the junction. The results from the five chalkland transects showed that, with the exception of fine materials, soil properties altered gradually along the catena, and that there was a diffuse transition zone from net erosion to net deposition in the surface soil.

The effect of hillslope curvature on soil properties was detected by David K. Swanson (1985) in a study of catenae in coarse-grained tills (Plates 7.2a, b). The tills formed two moraines near Willow Lake, in the Wind River Mountains, Wyoming. By plotting soil properties against slope curvature, it became evident that soils on convex slopes differ from soils on concave slopes along a catena (Figure 7.8). The effect of slope curvature was more

Table 7.3 Regression equation of soil properties against slope curvature

Soil property	Catenae	Coefficient of determination, R^2 (%)	Regression equation
Harden Index	Pinedale 1, 2, and 3	37.8	$y = 0.175 + 1.6468x$
	Bull Lake 1, 2, and 3	50.8	$y = 0.221 + 14.2495x$
Stone grus (%)[a]	Pinedale 2	8	$y = 21 + 238.47x$
	Bull Lake 2	31.7	$y = 48 + 4,245.67x$
Mean clay (%)[a]	Pinedale 2	87.4	$y = 6.1 + 115.80x$
	Bull Lake 2	100	$y = 8.3 + 517.03x$
Mean clay (g)[a]	Pinedale 2	86.1	$y = 9.3 + 181.22x$
	Bull Lake 2	77.3	$y = 13.3 + 1,128.49x$
Mean free aluminium (%)[b]	Pinedale 2	71.7	$y = 0.0469 + 0.4807x$
	Bull Lake 2	99.2	$y = 0.0636 + 4.0608x$
Free aluminium (g)[b]	Pinedale 2	44.8	$y = 0.0728 + 0.7169x$
	Bull Lake 2	95.7	$y = 0.1022 + 13.6356x$
Mean free iron (%)[b]	Pinedale 2	77.3	$y = 0.259 + 4.5114x$
	Bull Lake 2	61	$y = 0.515 + 15.6848x$
Mean free iron (g)[b]	Pinedale 2	70	$y = 0.425 + 8.5398x$
	Bull Lake 2	99.8	$y = 0.780 + 81.9585x$

Notes: [a]Till soils only
[b]Dithionite soluble
Source: After Swanson (1985)

marked on catenae formed in the 140,000-year-old Bull Lake moraine, than on the 20,000-year-old catenae formed in the Pinedale moraine; the difference presumably relates to the length of soil evolution. These age-influenced topographic relationships were found for a variety of soil properties (Table 7.3).

Some evidence suggests that catenary position is more important in understanding relationships between soils and topography than slope gradient. A study of morphological and selected chemical and physical properties of twelve pedons on three landscape elements (summit, shoulder, and backslope) was made on soils in north central Florida to evaluate the extent to which landscape position influenced soil genesis (Ovalles and Collins 1986). Thickness of A horizon, matrix colours with chroma greater than 2, percentage sand, pH, and organic carbon content below the A horizon were found to increase downslope (from summit, through shoulder, to backslope); mottles and matrix colours with chroma less than 2, silt and clay percentages, and total phosphorus content were found to decrease downslope. Chi-square tests suggested highly significant relationships between soil properties and soil-landscape position. Several relationships between soil properties, as measured by Spearman's rank correlation coefficient, ρ, were masked when computed using data for the whole landscape. In most cases, when $\rho < 0.5$ for the whole landscape, the correlation coefficients computed according to landscape position increased; when $\rho > 0.5$ for the whole landscape, most of the correlation coefficients computed for landscape positions decreased. This highlights the need to consider scale effects when studying geoecosystems.

A problem with establishing soil–slope relationships is that soils are highly variable over short distances. Soil-landscapes in Saskatchewan, Canada, can realistically be divided into three mapping units: convex upper slopes with shallow soils; concave lower slopes with deep soils; and depressional areas with gleyed soils (King *et al.* 1983). Smaller scale divisions were not possible owing to microscale variability of the soil. In a single 2.2 ha field in the North Carolina Piedmont, a colour-development index, used as an indicator of the degree of soil development, showed more than a threefold variation (Van Es *et al.* 1991). Two points are worth developing about the small-scale variability of soil (and landforms). First, the variations are not necessarily random, and measurements of a property taken close together are commonly more alike than those taken farther apart. This connectedness of the spatial variance structure of soil and landform data is presumably in essence a topographic influence on soil and landform evolution and can be investigated statistically (e.g. Selles *et al.* 1986) and stochastically (e.g. Kachanoski 1988). Second, the variation may result from chaotic dynamics in the soil and landscape (Phillips 1993a). These two ideas will be examined individually.

At a site lying 30 km east of Weyburn, Saskatchewan, soil cores were taken along a transect every metre in soils formed under native grassland (Kachanoski 1988). The soils had evolved in two glacial tills, the upper one about 20,000 years old and the lower one about 38,000 years old. The stratigraphic surface between the two tills was characterized by a sandy gravel layer lying between 1–2 m below the soil surface. Elevations were recorded around each sampling point at all nodes of a 3 m × 3 m grid, and microtopographic gradient and slope curvature were computed. The transect ran along a relatively gentle slope, the gradient of which was less than 0.5 per cent. Correlations between soil properties and microtopographic variables are shown in Table 7.4. The microtopographic variables do explain some of the soil properties, though the amount of variation explained is not high. Given the overall flatness of the terrain, it is perhaps surprising that any variation in soils properties is explained by topography. Multiple correlation between microtopographic variables and horizon thickness accounted for 25 per cent of the variance in A horizon thickness, and 9 per cent of the variance in B horizon thickness. Further investigations were carried out using power spectrum analysis and coherency estimates. The power spectra for slope curvature and A horizon mass are strikingly similar (Figure 7.9a) and show evidence of cycles at 7 m (0.14 cycles/m) and 3 m (0.32 cycles/m). Significant correlation between the two variables at the 7 m cycling frequency was established by coherency estimates. Power spectra for B horizon thickness

Table 7.4 Correlation between topography and soils properties along the Weyburn transect

		Microtopographic variables			
Soil property		Elevation	Slope gradient	Slope curvature	Sand lens depth
A horizon	Thickness	0.14	0.17	*0.38*	0.15
	Bulk density	0.18	*0.34*	0.12	−0.22
	Mass	0.07	0.27	*0.37*	0.08
	Total carbon	0.01	0.08	0.14	0.12
B horizon	Thickness	−0.08	−0.03	0.27	*−0.41*
	Bulk density	*−0.41*	*0.46*	−0.01	−0.03
	Mass	−0.16	0.06	0.28	−0.41
	Total carbon	−0.14	−0.06	−0.1	−0.38
Solum	Thickness	0.01	0.08	−0.03	*−0.3*
	Bulk density	*−0.05*	*0.5*	−0.17	*−0.35*
	Mass	−0.12	0.19	−0.07	*−0.35*
	Total carbon	−0.01	0.02	−0.03	−0.17

Note: Italicized correlations coefficients are significant at $p < 0.05$; emboldened and italicized correlation coefficients are significant at $p < 0.01$
Source: After Kachanoski (1988)

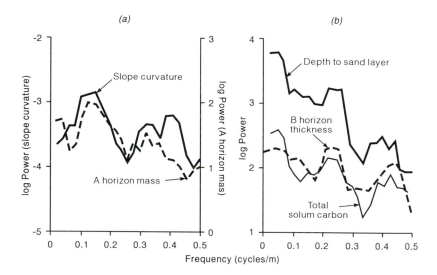

Figure 7.9 Power spectra for soils and topography at the Weyburn study site, Saskatchewan, Canada. (a) Comparison of spectra for A horizon mass and slope curvature. (b) Comparison of B horizon thickness, total solum carbon, and depth to sand layer
Source: After Kachanoski (1988)

and depth to the sand lens lying between the two tills also match one another, and both have a significant peak at 4.5 m (0.23 cycle/m) which is not found in the A horizon spectrum (Figure 7.9b). This suggests that processes influencing A horizon evolution do not have the same spatial variance relationships as those influencing the evolution of the B horizons. The indications are that A horizon variability is affected by surface curvature, acting through the redistribution of rainfall and the moisture in the root zone, and thus biomass production and leaching potential. But, at greater depths, the depth to the sand lens, which impedes the passage of water and increases water content above it, appears to be the main influence the redistribution of moisture (Kachanoski 1988). Indeed, the sand lens seems to have a long-term influence on moisture conditions. This is evidenced in the spectra for depth to the sand layer and total soil carbon, which are very similar. Total soil carbon is determined by vegetative growth, which in turn is influenced by moisture conditions.

By reinterpreting Jenny's 'clorpt' equation using non-linear dynamic systems theory, Jonathan D. Phillips (1993c) was able to show that the degree of soil-profile development, set in the context of the state-factor model, is asymptotically unstable and has a propensity for deterministic chaotic behaviour. An implication of this is that the spatial and temporal complexity of soil properties and soil evolution is inherent in the dynamics of the soil system, quite independently of any external stochastic forcing or original variations in environmental controls. Although it is not presently possible to isolate a chaotic attractor in field data where significant stochastic complexity is present, the extreme microscale variability in soil types and soil properties, in the apparent absence of parent material variations or other environmental influences, is consistent with chaotic behaviour in soil systems. This is because even minute differences in initial conditions may lead to divergent soil properties. As a field example of soil variability that seems to have arisen from almost identical initial conditions, Phillips used two closely related soil series on the North Carolina coastal plain. Properties of ten pedons in the Norfolk Series and ten pedons from the Goldsboro Series were used to establish an index of soil development. All the sites are on the lower coastal plain of North Carolina, on the same Pleistocene depositional sea-level terrace, formed in very similar moderately fine-textured coastal plain sediments, experience the same present-day climate, and support the same natural vegetation and land use. The two soils occur within the same part of the landscape and are part of a catenary sequence: the well-drained Norfolk Series occupies the edges of broad interfluves on upper valley-side slopes, and on convex upland ridges; the moderately well-drained Goldsboro Series occurs in slightly lower and wetter sites. Despite their being no significant differences in climate, parent material, topographic setting, biotic influences, or age of the pedons in each series, striking differences in the degree of profile development were found. The mean value of the development index for the

Norfolk Series was 13.6, and ranged between 6.26 and 18.31. For the Goldsboro Series, the mean development index was 10.15, and ranged between 3.23 and 16.02. In all the lower coastal plain soils, the development index ranges from approximately 1 to 22 (Phillips 1990, 1992). Such wide variations under seemingly almost identical environmental conditions strongly suggest sensitive dependence on initial conditions and evolution though chaotic dynamics (see also Phillips 1993a, 1993b).

Vegetation catenae

Milne's pioneering work on catenae, and the studies that followed in its wake, included changes of vegetation, as well as soils, over undulating topography. Vegetation catenae seem to have made something of a comeback in ecology, with many examples recently appearing in the literature. For instance, a detailed study was made of the terrain, soils, and vegetation in the R4D Research Site, Brooks Range Foothills, Alaska (D. A. Walker *et al.* 1989). A typical toposequence along the long west-facing slope of the research site is depicted in Figure 7.10. This catena is formed in till of Sagavanirktok age (mid-Pleistocene). The summit and shoulder have rocky mineral soils. This promotes a relatively high heat flux, thick active layers (>80 cm), and small amounts of interstitial ground-ice near the top of the permafrost table. Vegetation on these upper slopes is tussock-tundra. The lower hillslope elements – the lower backslope and footslope – have deep accumulations of *Sphagnum* peat overlying fine-grained colluvial deposits. The bog moss, *Sphagnum*, grows rapidly, forming a carpet that absorbs large quantities of water. This peaty mat insulates the ground, creating a shallow active layer (20–30 cm thick) and B horizons rich in ice. Paludification (the accumulation of peat) is most pronounced on footslopes where the vegetation is a moist or wet dwarf-shrub, moss tundra. On the lower backslope, a moist dwarf-shrub, tussock sedge (or non-tussock sedge) tundra occurs. Farther north, on the Arctic coastal plain, loess is the dominant parent material. An idealized alkaline-tundra toposequence formed in loess, and typical of catenae found around Prudhoe Bay near the Arctic coast, is shown in Figure 7.11 (D. A. Walker and Everett 1991). Eight common vegetation types and soils are identified. The eight vegetation stand types are grouped according to moisture characteristics into dry, moist, wet, and aquatic tundra. The toposequence is not normally associated with hillslopes. More commonly, it occurs in the patterns associated with ice-wedge polygons and small tundra streams. Indeed, microscale variations in topography are a strong control on vegetation and soils in the area.

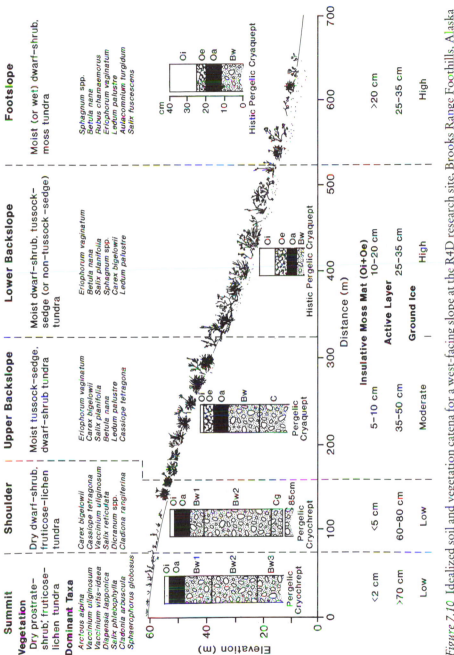

Figure 7.10 Idealized soil and vegetation catena for a west-facing slope at the R4D research site, Brooks Range Foothills, Alaska
Source: After D. A. Walker *et al.* (1989)

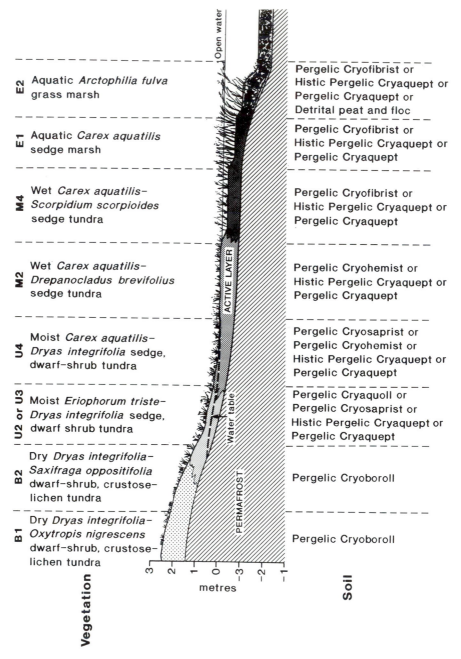

Vegetation

E2 Aquatic *Arctophilia fulva* grass marsh

E1 Aquatic *Carex aquatilis* sedge marsh

M4 Wet *Carex aquatilis–Scorpidium scorpioides* sedge tundra

M2 Wet *Carex aquatilis–Drepanocladus brevifolius* sedge tundra

U4 Moist *Carex aquatilis–Dryas integrifolia* sedge, dwarf-shrub tundra

U2 or U3 Moist *Eriophorum triste–Dryas integrifolia* sedge, dwarf shrub tundra

B2 Dry *Dryas integrifolia–Saxifraga oppositifolia* dwarf-shrub, crustose-lichen tundra

B1 Dry *Dryas integrifolia–Oxytropis nigrescens* dwarf-shrub, crustose-lichen tundra

Open water

ACTIVE LAYER

Water table

PERMAFROST

metres

Soil

Pergelic Cryofibrist or Histic Pergelic Cryaquept or Pergelic Cryaquept or Detrital peat and floc

Pergelic Cryofibrist or Histic Pergelic Cryaquept or Pergelic Cryaquept

Pergelic Cryofibrist or Histic Pergelic Cryaquept or Pergelic Cryaquept

Pergelic Cryohemist or Histic Pergelic Cryaquept or Pergelic Cryaquept

Pergelic Cryosaprist or Pergelic Cryohemist or Histic Pergelic Cryaquept or Pergelic Cryaquept

Pergelic Cryaquoll or Pergelic Cryosaprist or Histic Pergelic Cryaquept or Pergelic Cryaquept

Pergelic Cryoboroll

Pergelic Cryoboroll

Figure 7.11 Idealized soil and vegetation catena in alkaline tundra around Prudhoe Bay, north Alaska
Source: After D. A. Walker and Everett (1991)

SOIL LANDSCAPES

Three-dimensional topographic influences

According to the concept of soil-landscape systems, as presented by the author (Huggett 1975), the dispersion of all the debris of weathering – solids, colloids, and solutes – is, in a general and basic way, hugely influenced by land surface (and phreatic surface) form. It is organized in three dimensions within the framework imposed by the drainage network. In moving down slopes, weathering products tend to move at right angles to land-surface contours. Flowlines of material converge and diverge according to the curvature of the land-surface contours. The pattern of vergency influences the amounts of water, solutes, colloids, and clastic sediments held in store at different landscape positions. Of course, the movement of weathering products alters the topography, which in turn influences the movement of the weathering products – there is feedback between the two systems. If soil evolution involves the change of a three-dimensional mantle of material, it is reasonable to propose that the spatial pattern of many soil properties will reflect the three-dimensional topography of the land surface. This hypothesis can be examined empirically, by observation and statistical analysis, and theoretically, using mathematical models.

Investigating the effect of landscape setting on pedogenesis requires a characterization of topography in three dimensions. Early attempts to describe the three-dimensional character of topography was made by Andrew R. Aandahl (1948) and Frederick R. Troeh (1964). More recently, methods of terrain description have been explored by geographers and geomorphologists (e.g. Speight 1974; Dikau 1989; Moore *et al.* 1991). Topographic attributes that appear to be important are those that apply to a two-dimensional catena (elevation, slope, gradient, slope curvature, and slope length) plus those pertaining to three-dimensional landform (slope direction, contour curvature, and specific catchment area).

Edaphic properties

Three-dimensional topographic influences on soil properties were con-sidered in small drainage basins by the present author (Huggett 1973, 1975) and Willem J. Vreeken (1973), while André G. Roy and his colleagues (1980) considered soil–slope relationships within a drainage basin. Later work has confirmed that a three-dimensional topographic influence does exist, and that some soil properties are very sensitive to minor variations in the topographic field. A case in point is an investigation into natural nitrogen-15 abundance in plants and soils within the Birsay study area, southern Saskatchewan, Canada (Sutherland *et al.* 1993). Two sampling grids were used, each involving 144 points, in an irrigated field. The large grid was 110 × 110 m and the small grid 11 × 11 m (Figure 7.12). Samples of soils from both

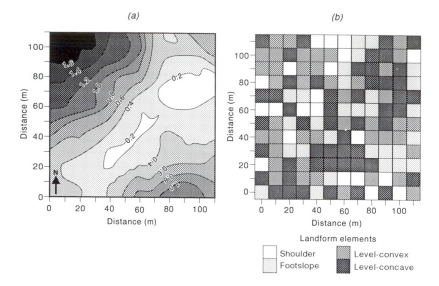

Figure 7.12 Study site, Birsay, Saskatchewan. (a) Topography. (b) Landform
elements in grid-cells
Source: After Sutherland *et al.* (1993)

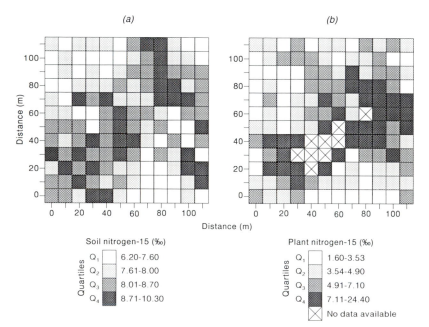

Figure 7.13 Quartile maps of (a) soil nitrogen-15 and (b) plant nitrogen-15 in the
study site at Birsay, Saskatchewan
Source: After Sutherland *et al.* (1993)

186

Table 7.5 Regression equation relating measured soil properties in the top 10 cm to significant terrain attributes ($p < 0.01$)

Soil properties	Intercept value	Slope	Topographic attributes				Coefficient of determination, R^2 (%)
			Wetness index	Stream power index	Aspect[a]	Profile curvature	
A horizon depth (m)	0.096	-0.053(1)[b]	0.031(2)	–	–	–	50
Organic matter (%)	0.285	–	0.190(1)	-0.070(2)	-0.002(3)	–	48
Extractable phosphorus (mg/kg)	-3.039	-1.466(1)	2.311(2)	-0.769(3)	–	–	48
pH	7.508	0.191(1)	–	–	-0.003(2)	–	41
Sand (%)	46.417	2.941(1)	-1.320(2)	–	–	27.592(3)	52
Silt (%)	23.466	-2.009(1)	2.076(2)	–	–	-27.241(3)	64

Notes: [a]Measured in degrees clockwise from west
[b]Numbers in parentheses indicate the order in which the variables were entered into the regressions
Source: After Moore et al. (1993)

grids, and samples of durum wheat (*Triticum durum*) from the large grid, were analysed for nitrogen-15 (Figure 7.13). Spatial statistical analysis indicated that the distribution of nitrogen-15 was random in the small grid, but in the large grid it was concentrated in depressions and followed the same pattern as denitrification activity and related soil properties (Eh, soil water content, bulk density, and total respiration). Spatial variability of nitrogen-15 in plants was greater than that in soils. Extreme outliers of nitrogen-15 in plants were associated with the landscape elements with highest denitrification activity and lowest Eh values. Elevation was the single most important variable for both plant and soil nitrogen-15 abundances, accounting for 26 per cent of the variation in soil nitrogen-15 and 31 per cent of the variation in plant nitrogen-15. Overall, the analysis suggests that topography had a significant influence on landscape patterns of nitrogen-15 in soil and plants.

Relationships between soil attributes and terrain attributes were revealed in a landscape at Sterling, Logan County, Colorado (Moore *et al.* 1993). The area of the site is 5.4 ha. Relative elevation and A horizon thickness were measured at, and soil samples were taken from, 231 locations on a 15.24 × 15.24 m grid. Several primary and secondary topographic attributes were derived from the elevation data. Primary attributes were slope (per cent), aspect (degrees clockwise from north), specific catchment area (m²/m), maximum flow-path length (m), profile curvature (/m), and plan curvature (/m). Secondary attributes were a wetness index, a stream-power index, and a sediment-transport index. The 'best' combination of terrain variables for explaining variation in soil attributes was explored using stepwise linear regression (Table 7.5). Slope and a wetness index (Figure 7.14) were the topographic variables most highly correlated with soil properties, accounting individually for about 50 per cent of the variability in A horizon thickness,

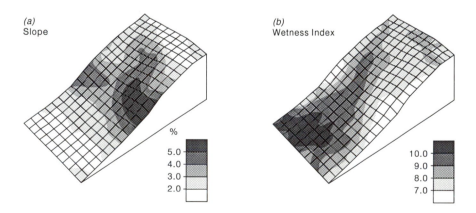

Figure 7.14 A catena at a site near Sterling, Colorado. (a) Slope. (b) Wetness index. The grid is 15.24 × 15.24 m
Source: After Moore *et al.* (1993)

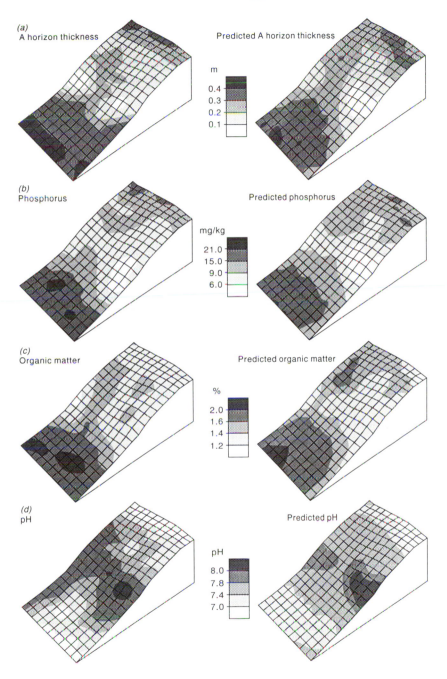

Figure 7.15 Measured and predicted soil attributes at the Sterling site
Source: After Moore *et al.* (1993)

organic matter content, pH, extractable phosphorus, and silt and sand contents. The regression equations were used to predict the spatial distribution of soil attributes (Figure 7.15). Correlation among the three terrain attributes and soil attributes suggests that pedogenesis in this landscape has been influenced by the way in which water flows through and over the soil.

Soil processes in landscapes may be modelled mathematically. An early and very elementary attempt to do this was made by the present author (e.g. Huggett 1975). A simple model, with a program listing, is described in Huggett (1993). More sophisticated models of the same kind have been developed to simulate nutrient dynamics within landscapes (e.g. Bartell and Brenkert 1991).

Regolith

Processes affecting regolith thickness operate in three dimensions and should, therefore, be influenced by landscape form as well as landscape position. Results of several studies imply a connection between land form and soil erosion. Topographic surveys of small plots (7–10 ha) in fields lying about 80 km north-west of Saskatoon, Canada, were carried out along transects using an average density of 50 observations per hectare (Pennock and de Jong 1987). Slope curvature, contour curvature, and slope gradient were calculated at all points. Soil samples were taken at 50-m intersections of a grid, and analysed for caesium-137, the redistribution of which compared to native or control sites was used as a measure of soil erosion. Distinct differences in soil gains and losses were associated with landform elements according to the vergency pattern of flowlines: on shoulders, convergence caused enhanced erosion; deposition was associated mainly with footslopes. This pattern is clear for the data for the entire data set (Figure 7.16). A study in a 10.5 ha first-order drainage basin, located in the south-western part of Bureau County, north-western Illinois, disclosed that erosion varies with landscape position (Kreznor et al. 1989). Transect data of all geomorphic units at a cultivated site showed that shoulders were slightly or moderately eroded while the lower backslopes and upper footslopes were either severely or very severely eroded, hinting that slope length was the key determinant of erosion. Landscape form also exerted an influence on erosion: geomorphic units with positive contour curvature (hollows) were less eroded than those with negative contour curvature (spurs).

The thickness of the debrisphere commonly varies according to present topography, but historical changes in landforms also exert a long-lasting effect. This appears to be so in the case of loess thickness at two sites in northern Delaware, United States (Rebertus et al. 1989). The loess lies upon a palaeosurface that is more irregular than the present land surface, and was probably a scoured erosional surface with occasional deeper concavities, possibly relict ice-wedge casts formed in a periglacial environment. Total

Figure 7.16 Soil gains and losses in different landform elements in small plots in
fields north-west of Saskatoon, Canada
Source: From data in Pennock and de Jong (1987)

relief of the two surfaces is similar at both sites, although slope lengths and
gradients differ between the sites. The palaeosurface seems to have received
an uneven covering of loess. Concavities and declivities received thicker
deposits than protuberances, though the thicker deposits in concavities might
have resulted from erosion of convexities and redeposition in lower
landscape positions. Loess deposition has smoothed the topography so that
the loess surface mirrors the general contours of the palaeosurface, but lacks
the irregularities. Three-dimensional influences on regolith thickness are also
seen in steepland hillslopes of Taranaki, New Zealand (DeRose *et al.* 1991).
Regolith depth varies greatly on both spur and swale sites, but the mean
depth for swales (122 cm) is significantly deeper than for spurs (59 cm), as is
the variation about the mean – the standard deviations are 79 cm for swales
and 43 cm for spurs. These differences point to major dissimilarities in the
processes influencing regolith depth on convex and concave sites. The mean
depth of regolith on spur profiles is independent of mean profile slope, but
decreases with increasing mean profile slope on swale profiles (Figure 7.17).
Regolith depth varied considerably along slope profiles, often varying as
much as 50 cm within a metre. The variations showed a basic cyclical pattern
related to changes in instantaneous slope. At any point on a hillslope, at least
in steeplands, the depth of regolith depends on several interacting factors
including: vegetation, erosional processes, topography (slope, curvature, and
position), hydrology, and external sources of soil material (loess and tephra).

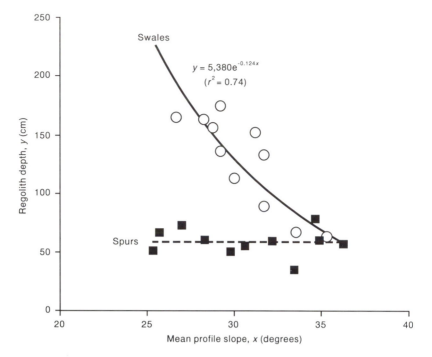

Figure 7.17 Relationship between mean regolith depth and mean profile slope for
slope profiles in swales and on spurs
Source: After DeRose *et al.* (1991)

Microtopographic variations in regolith thickness seem largely determined
by the effects of a previous forest cover, in which tree-throw produced a
hummocky land surface that has persisted after forest clearance. Along spur
profiles, regolith depth was greatest on sites of concave bedrock, shallowed
towards the stream at the slope base, and was thinnest where contour
curvature was extreme and there was a steep slope with neighbouring swale
profiles. Along swale profiles with mean slope angles less than 31°, regolith
depth was inversely related to instantaneous slope (shallowed with increasing
slope angle), and deepened in concave bedrock sites and towards the stream.
In swale profiles with mean angles more than 31°, regolith shallowed with
increasing slope angle but was not influenced by hillslope position or
curvature, and varied greatly in depth owing to repeated landsliding.

Brian T. Bunting (1965: 75) conjectured that position within a drainage
network should have a bearing on soil type and soil properties. This
relationship would be expected if, as seems reasonable, drainage basin
expansion and integration produce a sequence of valley development in
which each component drainage basin in the network has a characteristic
combination of landscape elements. In turn, the landscape elements influence

Table 7.6 Soil and slope characteristics according to stream order in a Queensland site

Stream order	Mean slope angle (°)	Convexity (°/hm)	Concavity (°/hm)	Basal channel type	Summit	Backslope	Toeslope
						Soil type	
1	9.4	16.4	37	Percoline	Deep red loam	Deep red loam	Deep red loam (gleyed)
2	16.9	30	41	Seepage line	Red loam	Red loam	Acid red loam (gleyed)
3	19	37	33	Poorly defined channel	Shallow loam	Skeletal loam	Deep red podzol
4	20	45	24	Well-defined channel in bedrock	Skeletal loam	Skeletal	Deep red podzol
5	21.2	42	44	Well-defined channel in bedrock and alluvium	Skeletal loam	Skeletal	Alluvial
6	16	40	51	Well-defined channel in bedrock and alluvium	Skeletal loam	Skeletal	Alluvial

Source: After Arnett and Conacher (1973)

the slope and soil processes. Thus the evolution of a soil profile should be influenced by the landscape at several scales: microscale landforms, hill-slopes, single drainage basins, and the entire landscape or drainage basin network. The relationship between soil, soil catenae, and stream order has not been much explored, but the available evidence suggests that it does exist. In the Rocksberg Basin, Queensland, soil type is related to catenary position and stream order (Table 7.6) (Arnett and Conacher 1973). A study conducted on soils and landscapes in drainage basins in central Spain suggests a structural correlation between the spatial organization of the fluvial systems and the soil landscape (Ibáñez *et al.* 1990). And John Gerrard (1988, 1990a), in an investigation of soils on Dartmoor, England, found that relationships between soil type and slope position were modulated by location within a drainage basin network (stream order).

An attempt to probe relationships between soil types and stream order was made by the present author (Huggett 1973; see also Warren and Cowie 1976), using soil maps. The method was to divide the soil-landscape under consideration into drainage basins of increasing order. Next, the cumulating area of soil types in each drainage basin is plotted against the cumulating drainage area, from first-order, headwater valleys to the largest-order basin in the area. The regularity of the curves so produced is striking, and suggests that they may reflect the influence of underlying physical processes linking soil type and its position within a hierarchy of drainage basins. Soil-landscapes in west Essex were found to exhibit three basic types of curve, the interpretation of which depends on the kind of parent material the soil is formed in – bedrock or alluvial and colluvial deposits (Figure 7.18) (Huggett 1973). In the Frieze Hall Catchment, the cumulating-area curve for soils of the Windsor Series, formed in London Clay, is 'concave' (Figure 7.18a): it has an 'area lag', not appearing until the drainage area is 0.15 km^2, then rises at an increasing rate with increasing drainage area. The curve for the Curdridge Series, formed in loamy Claygate Beds, is linear: it has a small 'area lag', and then rises to a constant value marking the point at which the river system has cut below the Claygate stratum. The curve for the Bursledon Series, formed in sandy Bagshot and Claygate beds, is 'convex': it rises at a decreasing rate with increasing drainage area. The small 'area lag' in all three cases arises because much of the bedrock on the summits and interfluves of headwater basins is covered by Pebbly Clay Drift, the parent material of the Essendon Series. Convex, linear, and concave curves are also produced by soils formed in alluvium and colluvium (Figure 7.18d to f). In the Fox Wood Catchment, the cumulating curve for the Titchfield Series, formed mainly in valley gravels, begins when the drainage area is about 0.45 km^2, then rises at an increasing rate. The curve for the Enbourne Series, formed in river alluvium, is linear in the Fox Wood Catchment, but in the nearby Beacon Hill Catchment, it is convex (Figures 7.18e and f). The 'area lag' for alluvial and colluvial soils results from a critical area being required for alluvium or

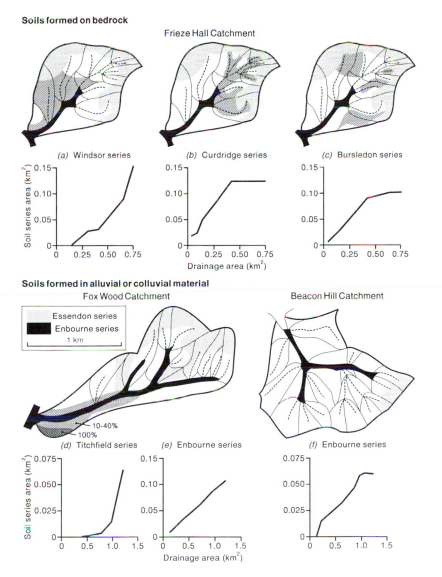

Figure 7.18 Area occupied by soil series versus drainage area for small catchments in west Essex, England. (a)–(c) Soils formed largely in bedrock. (d)–(f) Soils formed in superficial deposits
Source: After Huggett (1973)

colluvium to be collected and deposited.

Far more work on relationships between soils, soil properties, and stream order needs to be carried out before any firm conclusions can be drawn. However, there does appear to be a relationship between these variables. This fact vindicates the view that soil should be viewed as a three-dimensional body interacting with the landscape in which it evolves, and lending strong support to the concept of geoecosystems.

SUMMARY

Many processes in geoecosystems influence, and are influenced by, topographic fields. Relief, aspect, slope gradient, slope curvature, slope length, and contour curvature all have demonstrable effects on geoecosystems, particularly at microscales and mesoscales. Animals, plants, and soils adjust to the microclimates associated with slopes of differing aspect. The concatenation of slope soils by the downhill movement of water, solutes, colloids, and coarser material creates soil toposequences (soil catenae). Soil properties vary in a systematic way along a catena. These differences in soil properties lead to the evolution of a vegetation catena, with each slope element carrying a distinctive vegetation. The downhill flux of materials is modulated by the three-dimensional nature of topography. The result is that many soil-landscapes are organized in the framework of drainage basins. Several studies have shown that three-dimensional topographic effects are important in understanding the spatial pattern of soil properties, both in the edaphosphere and in the debrisphere. Some evidence suggests that position within a drainage network also accounts for some of the spatial pattern in soils.

FURTHER READING

There is a substantial literature on the relations between topography and ecosystems, though much of it is in journals. Good reviews of topography and soils are found in Birkeland's *Soils and Geomorphology* (1984), Jenny's *The Soil Resource: Origin and Behavior* (1980), and Gerrard's *Soil Geomorphology: An Integration of Pedology and Geomorphology* (1992). The relevant sections of Raymond B. Daniels and Richard D. Hammer's *Soil Geomorphology* (1992) are well worth perusing. Work on relationships between topography and vegetation is reported in ecological and botanical journals such as *Journal of Ecology* and *Ecology*. Early work on aspect and vegetation is noted in the paper by Cantlon in *Ecological Monographs* (1953). Several studies of vegetation catenae have appeared over the last ten years or so. Those referred to by D. A. Walker and his colleagues (1989, 1991) provide a lead into the literature. The influence of microclimates in general, and not just those related to topography, on animals and plants is discussed in *Microclimate, Vegetation and Fauna* (1992) by Flip Stoutjesdijk and Jan Barkman.

8

INSULARITY

Insularity is a topospheric feature that greatly influences geoecosystems. True islands form where submarine hills and mountains reach the sea or lake surface. They are typically considered to be smaller than continents, Greenland being the largest at present. Logically, continental land-masses are islands, albeit enormous ones. To be sure, Australia and Antarctica are island-continents, and South America was an island-continent from 65 to 3 million years ago. Whether designated islands or not, continents that have been isolated for a considerable time manifest insular features in their fauna and flora. At the other end of the spatial scale, a true island conventionally ceases to be an island when it cannot sustain a supply of fresh water; it is then simply a beach or sand bar. The critical area required to carry a stock of fresh water is about 10 ha. True islands are divided, on the basis of geology, into two broad groups: oceanic islands and land-bridge islands. Oceanic islands are all either of volcanic origin, or made of coral rock, or both. Land-bridge islands either lately formed part of a nearby mainland or else, even though recent separation cannot be proved, have a structure similar to continental lands. They were formerly called continental islands. Useful though this classification be, it fails to capture the infinite variety of details displayed by oceanic and land-bridge islands.

As well as these two main types of island, there are two minor types: islands in rivers and lakes and habitat islands. Islands in rivers and lakes are formed by the direct accumulation of marine or fluvial sediment. They do not occupy a great area on a global scale, and are usually parts of deltas or estuary fillings, or unusually large coastal bars and sand-banks. Most natural habitats, not just truly insular ones, also possess a degree of insularity – streams, caves, gallery forest, tide pools, taiga as it breaks up into tundra, and tundra as it breaks up in taiga, mountain peaks, nature reserves, cities, parks within cities. These are habitat islands, and may be much smaller than the critical 10 ha required to qualify as a true island.

True islands are relatively isolated places that enjoy a maritime climate. Compared with the mainland, they are windier but subjected to lesser extremes of temperature. Many of them are vulnerable to severe meteorological and hydrological events – hurricanes, storms, and tsunamis – that

cause disturbance of island biotas. These characteristically insular environmental conditions create idiosyncratic features of island landscapes, and
notably of island life. Owing to its individuality, life on islands provides
useful clues to evolutionary and biogeographical processes, so it is not
surprising that islands are seedbeds of ideas in biology and biogeography
(MacArthur and Wilson 1967). The effects of insularity on life will now be
considered, species and communities being taken separately. The salient
point that shines through the discussion concerns the responsiveness and
adaptability of populations and communities to biotic and abiotic environmental constraints: the slight variations of phenotype fashioned by the
genealogical hierarchy are soon sorted and sifted by environmental factors to
produce species nicely tuned to the circumstances in which they live; and,
similarly, the constant turnover of species in a community strives towards a
steady state that is constantly shifting in response to environmental change.

ISLAND SPECIES

Natural selection and the origin of species

Life on islands gave clues to Charles Robert Darwin and Alfred Russel
Wallace that led them to the idea of natural selection as a mechanism
explaining the evolution of species (e.g. Darwin 1859; Wallace 1880). Today,
islands are still aiding studies on natural selection. Anita Malhotra and Roger
S. Thorpe (1991b) believe that they have demonstrated selection in action by
manipulating natural populations of the Dominican lizard (*Anolis oculatus*)
(Plate 8.1). Previous work had shown that complex patterns of geographical
variation in the morphology of the lizard (body size and shape, colour
pattern, and scalation) correlated, both univariately and multivariately, with
the considerable altitudinal and longitudinal variation of climate and
vegetation on Dominica (Malhotra and Thorpe 1991a). Malhotra and Thorpe
translocated several ecotypes of the species into large experimental enclo-
sures, and monitored them over two months. It was found that the
magnitude of the difference in multivariate morphology between the
survivors and non-survivors within each enclosure correlated with the
magnitude of the difference between the ecological conditions of the
enclosure site and the original habitat, and similar relationships were found
for three indices of fitness of the survivors.

Several classic examples of evolutionary processes come from islands.
Adaptive radiation is clearly seen in Darwin's finches (Geospizinae) on the
Galápagos Islands, the honeycreepers (Drepanidae) on Hawaii, and the
marsupials in Australia. Insularity seems to influence the dynamics of
evolution, particularly the extinction rate. Natural extinction on islands is a
complex process that may be linked to the taxon cycle (Ricklefs and Cox
1972, 1978). The rate of extinction on islands does appear to be high, though

Plate 8.1 A Dominican lizard (*Anolis oculatus*), Atlantic coast ecotype, male.
Photograph by Anita Malhotra

much of it may be attributed to human interference. Island populations are vulnerable to all introductions. This is the case for avian extinctions in the period 1680 to 1964 (Thomson 1964). During this time, 127 races or species of bird became extinct. Of these, eleven occurred on continents, twenty-nine on large islands, and eighty-seven on small islands. The significance of these data becomes crystal-clear when it is noted that only one-tenth of the world's avifauna inhabits islands. Most of the avian extinctions were caused by humans, but natural extinctions have occurred. In the Shetland Islands, the song thrush (*Turdus philomelos*) was absent during the nineteenth century. A breeding population established itself in 1906. By the 1940s there were about twenty-four breeding pairs. After the severe winter of 1946–47, a mere three or four pairs remained (Venables and Venables 1955). And, between 1953 and 1969 the song thrush population became extinct.

Despite their having high rates of extinction, many islands house species that have become extinct elsewhere; these species are referred to as relics. Island-continents and very large islands generally deliver the best examples of relict species: the marsupials in Australia (though a few marsupials still live in South America and southern North America); the lemur family (lemurs, indri-lemurs, and aye-ayes) in Madagascar and the Comoro Islands (E. P. Walker 1968); and the tuatara (*Sphenodon*), a primitive lizard-like reptile and the only living member of the order Rhynchocephalia, in New Zealand (Crook 1975). Many smaller islands also harbour relics, though they are less well known. The shrew-like insectivores *Nesophontes* and *Solenodon* on the Great Antilles are examples (MacFadden 1980). Oceanic islands commonly hold forms found nowhere else – endemics. On islands of the North Atlantic Ocean, the number of avian subspecies, expressed as percentages of the avifauna, is as follows: Ireland 3 per cent; Iceland 21 per cent; the Azores 30 per cent; and the Canaries 45 per cent (Lack 1969). On Tristan da Cunha, in the South Atlantic, all land bird species are endemic. The remoter islands in the Pacific ocean have endemic subfamilies and families of birds – the Hawaiian honeycreepers and Darwin's finches are well-known examples.

Geographical clines are found on islands, even on relatively small ones. Clinal variation on small islands is often called microgeographical variation. Plainly, the zoologists' definition of micro does not accord with the landscape ecologists' or geomorphologists' definition – by the criteria established earlier in the book, clinal variation on islands is mesogeographical variation, but what an ugly word that is (almost as ugly as Mesoamerica). Considerable clinal variation in colour pattern, scalation, and body dimensions is displayed by the skink, *Chalcides sexlineatus* (Plate 8.2), endemic to the island of Gran Canaria, which lies to the east of Tenerife in the Canary Island archipelago (R. P. Brown and Thorpe 1991a, 1991b). As this study uses multivariate methods to test rival hypotheses of microgeographical variation, and is thus in line with the approach encouraged in the present book, it will be described in detail.

Plate 8.2 Dorsal view of a Gran Canarian skink (*Chalcides sexlineatus*) from the
north of the island.
Photograph by Richard P. Brown

Gran Canaria has an area of 1,523 km^2 and a highest elevation of 1,949 m,
but its landscape is heterogeneous in terms of climate and vegetation which
are roughly zoned north to south (Figure 8.1). In total, 316 male and 375
female *Chalcides* were studied from forty-seven localities. Five linear body
dimensions and four scalation characters were taken on each specimen.
Geographical variations in body dimensions were described using multiple
group principle component analysis (MGPCA), canonical variate analysis
(CVA), analysis of variance (ANOVA), and analysis of covariance
(ANCOVA). Independence of each characteristic of scalation was tested for
by pooled within-group correlation (group meaning sex in this case).
Geographical variation and sexual dimorphism in each character of scalation
was tested using two-way ANOVA for both groups, while the generalized
geographical variation in scalation was investigated using CVA. Congruence
in the patterns of geographical variations in body dimensions and scalation
were tested by computing appropriate product–moment correlation coefficients. The results for body dimensions showed clines following a north-east
to south-west gradient, save snout-vent length which displayed a mosaic
pattern (Figure 8.2). The fifth MGPC represents the smallest percentage of
the total within-group variation (0.2 per cent in males and 0.4 per cent in
females) but the largest among-group variation (57.8 per cent in males and

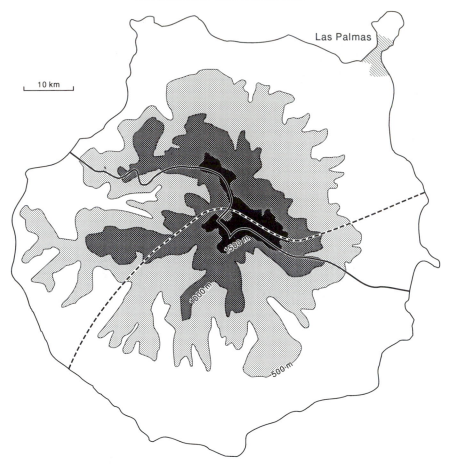

Figure 8.1 Gran Canaria: topography (elevation in m); dividing line between northern region of lush vegetation and arid southern region of sparse vegetation (unbroken line); and putative division between two species of *Chalcides sexlineatus* (broken line)
Source: After R. S. Brown and Thorpe (1991a)

67.7 per cent in females). It is a head-length factor and gives the clearest pattern of microgeographical variation: a stepped cline running from the highest scores (shortest heads) in the south-west of the island to the lowest scores (longest heads) on the mid-altitudes of northern slopes (Figure 8.3). Variation patterns in scalation are shown in Figure 8.4. Evidence of a step in the cline is present, the values for frenocular scales displaying the steepest transition zone of all the characters studied. Generalized scalation and body dimensions are congruent ($r = 0.79$, $p < 0.001$ for males, and $r = 0.77$, $p < 0,001$ for females).

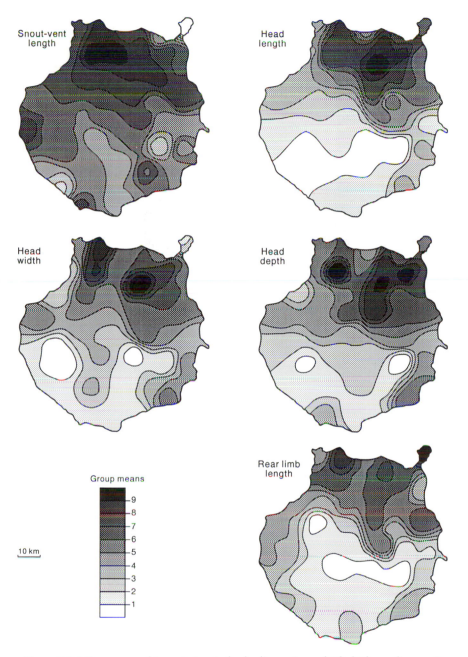

Figure 8.2 Microgeographic variation in body dimensions of *Chalcides sexlineatus* in Gran Canaria. The data are male group means scaled from 0 to 10. Male and female among-locality variation was congruent for each body dimension
Source: After R. S. Brown and Thorpe (1991a)

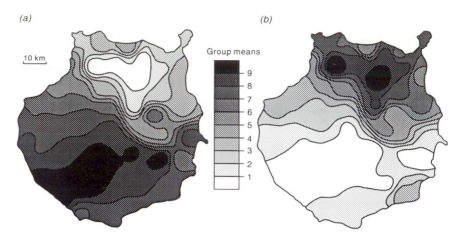

Figure 8.3 Microgeographic variation of (a) body dimensions and (b) generalized body dimensions as revealed by the fifth axis of a multiple group principal component analysis (MGPCA). The data are male group means scaled from 0 to 10. Male and female among-locality variation was congruent for MGPCA axis 5
Source: After R. S. Brown and Thorpe (1991a)

To explain the microgeographic variation in *Chalcides sexlineatus* on Gran Canaria, four hypotheses were proposed: a climate–vegetation hypothesis, a climate–vegetation plus gene-flow hypothesis, an altitude hypothesis, and a two-species hypothesis. The last of these derives from a suggestion in an earlier study that *Chalcides sexlineatus* is present in the southern and eastern parts of Gran Canaria, whilst '*Chalcides bistriatus* species complex' is found in the northern and western parts of the island. The climate–vegetation hypothesis is based on the fact that, climatically, the island may be divided into two (Plates 8.3a, b): the north-facing slopes in the north half of the island are much cloudier (have fewer sunshine hours), colder, and receive more rainfall than the south-facing slopes in the south half of the island; and that, in consequence, there is a sharp divide between the lush vegetation of the northern slopes and the sparse vegetation of the arid southern slopes (Figure 8.1). The second hypothesis included gene flow between the northern and southern categories defined in the first hypothesis by scoring localities according to their perpendicular distance from an axis running north-northeast to south-southwest. The altitudinal hypothesis took account of the fact that the ecological variation contains an altitudinal element, the seasonality of the climate and diurnal temperature range becoming greater with increasing elevation. The hypotheses were tested by partial correlation and by Mantel tests. An advantage of Mantel tests over partial correlation analysis is that they can compare multidimensional matrices. In all, fourteen male and fourteen female partial correlation analyses were computed

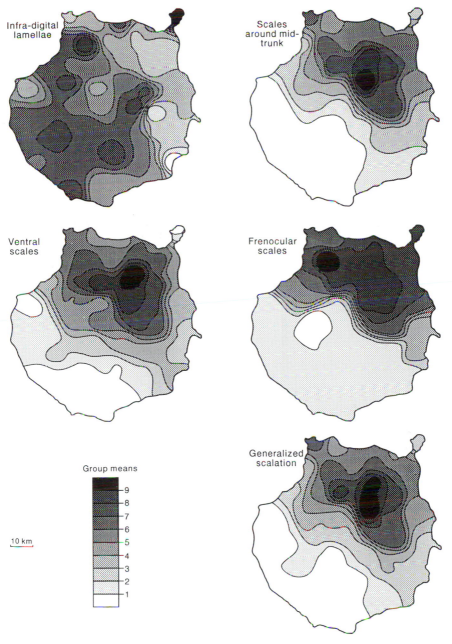

Figure 8.4 Microgeographic variation in scalation of *Chalcides sexlineatus* in Gran Canaria. The data are male group means scaled from 0 to 10. Male and female among-locality variation was congruent for each scalation character
Source: After R. S. Brown and Thorpe (1991a)

205

Plate 8.3a Gran Canaria, an island of contrasting vegetation types: the lush north.
Photograph by Richard P. Brown

Plate 8.3b Gran Canaria, an island of contrasting vegetation types: the arid south.
Photograph by Richard P. Brown

between body dimension characters and scalation and scores representing the four hypotheses. The results rejected the hypothesis that there are two species of *Chalcides* on Gran Canaria, the altitude hypothesis for body dimensions (except for rear limb length), and the climate–vegetation hypothesis for body dimensions, but upheld the climate–vegetation plus gene-flow hypothesis for body dimensions. The climate–vegetation hypothesis and the altitude hypothesis were not rejected as causes of generalized geographical variation in scalation. Mantel tests rejected the altitude hypothesis as a possible cause of the geographical variation in body dimensions, but did not reject the other three hypotheses; for scalation, they did not reject any of the hypotheses save for the altitude hypothesis in the case of female scalation. The single Mantel tests rejected fewer null hypotheses than the partial correlation analyses, a fact which underscores the importance of practising simultaneous hypothesis testing when trying to differentiate between several statistically non-independent hypotheses (R. P. Brown and Thorpe 1991a: 59).

The north–south pattern of microgeographical variation of *Chalcides sexlineatus* and its close relationship with current ecological conditions mirrors the microevolution of the lizard *Gallotia galloti* on Tenerife (Thorpe and Brown 1989). The colour pattern of *Gallotia* correlates with the lush–arid aspects on Tenerife, and its scalation displays a similar combination of latitudinal (climate and vegetation) and altitudinal elements. Similarly, the colour pattern of *Chalcides viridans*, a sister species of *C. sexlineatus* living on Tenerife, adheres to the north–south variation, and populations from the south of both islands have conspicuous dorsal tail coloration (R. P. Brown *et al.* 1991). Such congruence of microgeographical variations among the species suggests that current selection pressures from the environment, rather than phylogenetic causes, are largely responsible for the intraspecific variation within the islands.

Dwarfism and gigantism

Endemic island species tend to be either much smaller or much bigger than their mainland relatives. Dwarf forms are common among animals. Sherwin Carlquist (1965: 174–175) gives several examples including a diminutive West Indian gecko (*Sphaerodactylus elegans*), very common on Cuba, which is a mere 34 mm long; and a rattlesnake living on Tortuga Island in the Gulf of California (*Crotalus tortugensis*) which is about 1 m long, nearly half a metre shorter than its mainland relative, *Crotalus atrox*. Pygmy fossil forms exist, the most remarkable of which are the dwarf elephants and hippopotamuses. A pygmy hippopotamus is known from Madagascar. Fossil elephants, many of them pygmy forms, are found on many islands: San Miguel, Santa Rosa, and Santa Cruz, all off the Californian coast; Miyako and Okinawa, both off China; and Sardinia, Sicily, Malta, Delos, Naxos, Serifos, Tilos, Rhodes,

Crete, and Cyprus, all in the Mediterranean Sea (Sondaar 1976; D. L. Johnson 1980).

Giant forms are found in many animal groups and in plants. The largest earwig in the world, measuring nearly 8 cm in length, lives on St Helena (Olson 1975). The Komodo dragon, found on the Lesser Sunda Island of Indonesia, is the largest living lizard. Males are about 3 m long. Island geckos are commonly twice the size of their continental cousins. The renowned Galápagos tortoise (*Geochelone elephantopus*) and its cousin, *Geochelone gigantea*, on Aldabra, are giants among tortoises. The larger Australian marsupials are veritable giants compared with their North and South American relatives. Many fossil giants are known. Those in Australia include Pleistocene giant marsupials, even more gigantic than those living today, such as *Diprotodon*, the size of a hippopotamus (Archer 1981). Madagascar, as well as supporting a range of living lemurs, carries fossil forms, all of which are larger than their living relatives (e.g. Carlquist 1965). It was also home for the largest bird that ever lived, the elephant bird (*Aepyornis maximus*) which weighed in at nearly half a tonne. In the plant kingdom, long-distance, jump dispersal favours non-woody species – herbs. Island environments offer opportunities for herbs with aspirations towards treehood. Many examples of 'giant' herbs are known (e.g. Carlquist 1965). The Juan Fernandez Islands, west of Chile, have fostered some remarkable tree-lettuces and shrub-lettuces. On St Helena live a group of trees belonging to the sunflower family (Compositae) whose mainland relatives are small herbaceous plants.

The response of body size to life on islands is complex. In mammals, some orders tend to evolve bigger forms on islands, some tend to evolve smaller forms (Table 8.1). As a rule, small mammals tend to evolve to larger size and large mammals tend to evolve to smaller size, and medium-sized mammals show no particular trend in size change (Van Valen 1973). An indication of

Table 8.1 The size of mammalian insular species[a]

Taxonomic group	Size of insular species compared with mainland relatives		
	Smaller	Same	Larger
Marsupials	0	1	3
Insectivores	4	4	1
Lagomorphs	6	1	1
Rodents	6	3	60
Carnivores	13	1	1
Artiodactyls	9	2	0

Note: [a]Based on a survey of 116 insular races or species mostly living on the islands off western North America and Europe
Source: After Foster (1964)

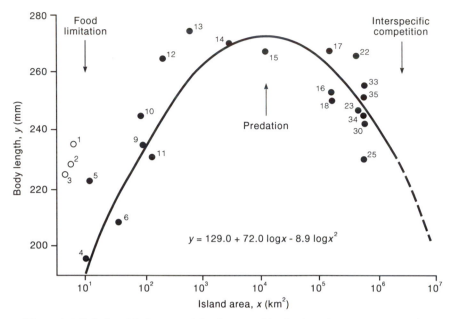

$$y = 129.0 + 72.0 \log x - 8.9 \log x^2$$

Figure 8.5 Relationship between island area and body size of Prevost's squirrel (*Callosciurus prevosti*), and factors deemed important in influencing body size, in Southeast Asia. Each numbered dot corresponds to the mean body size of a recognized subspecies, the numbers identifying particular islands (see Heaney 1978). Open circles show subspecies with small sample sizes (<4) and excluded from the statistical analyses
Source: After Heaney (1978)

the complexity of size change on islands was revealed by a study of the tricoloured squirrel (*Callosciurus prevosti*) in islands of Southeast Asia (Heaney 1978). Body length is small on islands with areas less than about 50 km², medium on islands with areas of about 100 km² and between 100,000 to 1,000,000 km², and high on islands of intermediate area, between about 200 to 100,000 km² (Figure 8.5). The size variations appear to have evolved in response to different pressures associated with island size. Small forms of squirrel have evolved on small islands in the face of food limitations, and on large islands because of competition with other species. Large forms have evolved on medium (and some large) islands where, confronted with predators, it pays to be big. More generally, John Damuth (1993) argues that a mammalian species living on an island free of competitors and predators will tend to evolve towards a body size of 1 kg (the optimum size for energy acquisition). This will be especially true of species that start out either big or small, and so have a greater adaptive advantage in changing size.

Dispersal

The dispersal of organisms has long puzzled biogeographers. Early explanations of why species live where they do focused on the restocking of the Earth by a pair of each species from Mount Ararat after disembarking from Noah's Ark (Browne 1983). Dispersal has been extensively studied since Charles Darwin and Alfred Russel Wallace did their pioneering work in the late nineteenth century. Insularity presumably affects dispersal because an organism must be able to make the journey from one favourable habitat to another across hostile terrain. Island colonization must involve jump dispersal or island-hopping – that is, the rapid passage of individuals over largish tracts of normally inhospitable terrain. In the case of true islands, the journey is made over the sea. An island-hop takes less time than the life-span of the individual involved. Species that are readily dispersed by wind, by floating, by flying, by being carried by flying animals, or by swimming are more likely to arrive on islands by 'hopping'. The process is certainly effective, witness the colonization of South America some 40 million years ago by rodents island-hopping from North America across an archipelago that temporarily joined the two continents. In view of the vast array of caviomorph rodents descended from the island-hoppers and seen in South America today, the process was plainly a small hop for a rodent but a gigantic leap for rodentkind.

There is a substantial body of observational evidence for island-hopping. It is known, for example, that new islands are rapidly colonized after their emergence. After the explosion of Krakatoa in 1883, a series of islets were produced some 40 km from Java and Sumatra. The recolonization of the islets by animals and plants is well documented. By 1934, there were 271 species present (Docters van Leeuwen 1936). Of these, 112 had been carried by wind, seventy-six by ocean currents, sixty-eight by animals (mainly birds), and fifteen by humans. Recent expeditions have recorded thirty species of land birds, nine bats, the Indonesian field rat, that indefatigable colonist the black rat, nine reptiles, including the monitor lizard (*Varanus salvator*), and over 600 invertebrates (Thornton and New 1988; Rawlinson *et al.* 1990). Another line of direct evidence for long-distance dispersal is the list of stray visitors to the Hawaiian Islands. Most of these unusual arrivals are not accustomed to prolonged flights – the osprey, the Arctic tern, the belted kingfisher, and many others. Figures for the widest ocean gap known to have been crossed by different animals is also revealing. The premier league comprises bats, land birds, insects, spiders, and land molluscs. These have managed to traverse 2,000 miles of ocean. Lizards have achieved half that distance. Tortoises and rodents have covered about 500 miles, and small carnivores about 200. The bottom division is occupied by large mammals and freshwater fish who have managed to cross a few tens of miles. Actually, some large mammals are proficient swimmers. Elephants, contrary to popular opinion, are excellent swimmers (D. L. Johnson 1980). They swim in a lunging, porpoise-like fashion, using their trunks as snorkels. They have been observed to reach a maximum of 2.7 km per hour and to cover a maximum distance of 48 km. It is not entirely clear why they should choose to take such long natatorial excursions, but the smell of food coming from offshore islands that they can see might be an inducement. This finding is important because it raises new possibilities for explaining the distribution of fossil elephants. Before this report on the proficiency of elephants as swimmers, it was widely assumed that elephants must have walked across former land bridges to the islands on which their fossil remains are found (though vicariance events are also a possibility). As it now seems feasible that the elephants could have swum to the islands, new explanations for the colonization of the islands are required.

Species that can disperse across ocean water with ease and, on arrival at an island, reproduce very rapidly have acquired a special name – supertramps. These aces among dispersers were first recognized by Jared M. Diamond (1974) on the island of Long, which lies off New Guinea (and is not to be confused with Long Island, New York). Long was devegetated and defaunated about two centuries ago by a volcanic explosion. The density of bird species on the island is currently far greater than would be expected. Of forty-three species, just nine accounted for the high avian density on the island. These nine are all supertramps. They specialize in occupying islands that are too small to sustain long-lasting, stable populations, or islands devastated by volcanic eruption, tidal wave, or hurricane. They are eventually ousted from these islands by competitors that can exploit resources more efficiently and that can survive at lower resource abundances.

At first sight, it is surprising that some islands house flightless forms of birds and insects, and plants that have lost their ability to disperse by wind and other means. On the Hawaiian Islands, several groups of insects have flightless representatives including butterflies and moths, flies, bees and wasps, ants, true bugs, and grasshoppers. A possible reason for the evolution of these flightless forms occurred to Charles Darwin: flighted forms of insects run a greater risk of being blown out to sea, and islands tend to be windy places. Flightless forms of birds are widely distributed among the world's islands. The Rallidae (rails, coots, gallinules) are well represented in the list of flightless birds: in the Pacific Ocean there are nineteen flightless forms, in the Atlantic Ocean three, and in the Indian Ocean one; many fossil forms are also known. Other flightless birds residing (or which have resided) on islands include penguins, cassowaries, emus, kiwis, moas, elephant birds, the flightless cormorant, the flightless duck, the steamer duck, the mesite, the kagu, the great auk, dodos, the kakapo, and the Stephen Island wren (before the last individual was eaten by the lighthouse keeper's cat). The dodo (*Raphus cucullatus*) is perhaps the best-known example of a flightless island resident, having been immortalized in the phrase 'dead as a dodo'. It probably evolved from African fruit pigeons of the genus *Treron* that became stranded on the island of Mauritius. In the absence of predators and with a surfeit of food available, its ancestors lost the ability to fly and, no longer needing to watch their weight, became big birds – flightless giants, though now believed to be sleeker and more fleet-footed than seventeenth-century paintings suggest (Kitchener 1993). The demise of the dodo in the 1680s was caused by the rats, pigs, and monkeys that arrived with sailors and that ravaged the dodos' ground nests. Explanations for flightlessness in birds are complex and beyond the scope of the present book.

ISLAND COMMUNITIES

Species–area relations

Large islands normally house more species than small islands. This is called the area effect. Related to it is the distance effect: remote islands hold fewer species than more accessible islands. These two effects reflect geographical influences on species diversity. They were recognized by naturalists in the eighteenth century. Eberhardt Zimmerman (1777) and Johann Reinhold Forster both noticed that the number of species found on continents and off-lying islands decreases in inverse proportion to the distance between an island and its neighbouring continent. Forster reasoned that this results from the physical features of islands becoming progressively dissimilar to the physical features of the continents the further they lay offshore. In particular, the climate of islands is always more equable than the climate of continental land-masses. Also, oceanic islands are commonly of volcanic origin and composed mostly of rocks different to those underlying continents. Physical differences such as these discourage a precise match of species between continents and islands. Forster (1778) opined that the environment plays a cardinal role in regulating the patterns of the organic world, and held that the number of species in an area is proportional to the available physical resources. In his own words, 'Islands only produce a greater or lesser number of species, as their circumference is more or less extensive' (Forster 1778: 169).

The area effect and distance effect were discussed by Alfred Russel Wallace (1869), Olof Arrhenius (1921), and Herbert A. Gleason (1922, 1925). They are both now widely observed in many island groups. Normally, the relationship is linear if number of species, S, and island area, A, are plotted on logarithmic co-ordinates. The best-fit line is described by the equation

$S = cA^z$

where c is the intercept value (the number of species expected to inhabit an island of unit area) and z is the rate of increase of species numbers per unit increase in island area. For amphibians and reptiles (the herpetofauna) living

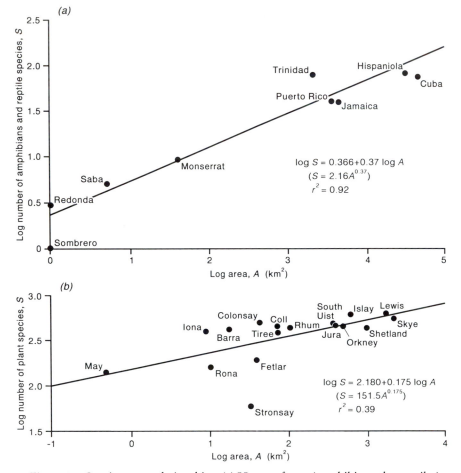

Figure 8.6 Species–area relationships. (a) Herpetofauna (amphibians plus reptiles) on some West Indian islands. The equation for this classic best-fit line differs from that presented in many books and papers because Trinidad and the tiny island of Sombrero were included in the computations. (b) Plant species on Scottish islands

Source: (a) from data in Darlington (1957) and (b) from data in M. P. Johnson and Simberloff (1974)

on the West Indian islands listed by Philip J. Darlington (1957: 483), the relationship, as depicted on Figure 8.6a, is

$$S = 2.16A^{0.37}$$

The line described by this equation fits the data for Sombrero, Redonda, Saba, Monserrat, Puerto Rico, Jamaica, Hispaniola, and Cuba very well. Trinidad lies well above the line. It probably carries more species than would be expected because it was joined to South America 10,000 years ago. For plant species on a selection of Scottish islands, the relationship (Figure 8.6b) is

$$S = 151.51A^{0.18}$$

A large number of studies have established the validity of this kind of relationship. As a general rule, the number of species living on islands doubles when habitat area increases by a factor of ten. For islands, values of the exponent z normally range between 0.24 and 0.34. In mainland areas, if the number of species is counted for areas of increasing size and a log–log line fitted, the z values fall within the range 0.12 to 0.17. This means that small areas on the mainland contain almost as many species as large areas on the mainland; this is not true of islands where smaller islands contain fewer species than large islands. The difference may be partly attributable to the distance effect, which expresses itself more strenuously in the case of islands.

A crucial question is why there is such a good relationship between island area and the diversity of fauna and flora. The answer is not simple. Two chief hypotheses have emerged (Gorman 1979: 24; Kohn and Walsh 1994). First, there is the habitat diversity hypothesis. In brief, this states that larger islands have more habitats and therefore more species. On the Galápagos Islands, 73 per cent of the variation in the number of bird species living on the different islands may be explained by plant species diversity (Abbott et al. 1977). In the Great Basin region of the United States, mountain peaks are 'islands' of boreal habitats. Ninety-one per cent of the number of boreal bird species occupying the peaks is explained by an index of habitat diversity measuring the extent and complexity of coniferous forest, riparian woodland, wet meadow, and aquatic habitat (N. K. Johnson 1975; see also Kratter 1992). Second, there is the area-alone hypothesis. Evidence exists suggesting that, irrespective of ecological diversity and habitat diversity, large islands contain more species than small ones. Parks within the city of Cincinnati are 'islands' in a sea of concrete and tarmac (Faeth and Kane 1978). Ninety-one per cent of the variation in the number of Dipteran species living in different parks is accounted for by park area alone. An index of habitat diversity made no significant contribution to the explained variance.

It is difficult to test the two hypotheses because, for most islands, area and habitat diversity are interrelated and it is difficult to unravel their effects. Some success in testing the hypotheses was made by Daniel S. Simberloff (1976). Pure stands of the mangrove (*Rhizophora mangle*) form little islands in the Bay of Florida. These islands support arthropods, insects, isopods, spiders, and scorpions. By reducing the size of three of the islands with power-saws over two successive years, and by setting up a third island as a control, Simberloff could observe the effect of reducing island size on species diversity. In all three cases, the species diversity fell in successive years, while that of the control island rose a little. The drop in diversity must have resulted from the diminishment of area, as the habitat diversity remained unchanged.

Table 8.2 Multiple regressions relating angiosperm, bird, and mammal species richness independently to island area and measures of island energy per unit area[a]

Species	Model	Adjusted coefficient of determination, R^2 (per cent)	Sample size, n
Angiosperms	$\log S = 4.8 + 0.41 \log A + 1.1 \log AET$	89	24
	$\log S = 2.1 + 0.63 \log AET \cdot A$	66	24
Birds	$\log S = -6.0 + 0.37 \log A + 0.69 \log PP$	90	28
	$\log S = -1.7 + 0.47 \log PP \cdot A$	79	28
All mammals	$\log S = 1.3 + 0.29 \log A + 0.42 \log PET$	72	161
Small mammals[b]	$\log S = 1.4 + 0.26 \log A + 0.45 \log PET$	66	155
Herbivores	$\log S = 1.2 + 0.16 \log A + 0.44 \log AET$	61	68
Carnivores	$\log S = 0.28 + 0.17 \log A + 0.19 \log AET$	65	32

Notes: [a]All relationships are significant at $p < 0.0001$ save carnivore species diversity versus AET for which $p = 0.2144$. AET is actual evapotranspiration, PET is potential evapotranspiration, and PP is net primary production
[b]Mammals weighing less than 5 kg
Source: After Wylie and Currie (1993)

A recent study showed that the influence of island area on species richness may not be as important as the influence of island energy. John L. Wylie and Currie (1993a) examined patterns of angiosperm, bird, and mammal species richness on islands using the multiple regression model

$$\log S = a + b \log A + c \log G$$

where S is species richness, A is island area, G is energy per unit area of island, and a, b, and c are regression coefficients. They measured energy using several variables – solar radiation, potential evapotranspiration, actual evapotranspiration, net primary productivity, and latitude – on a sample of land-bridge islands with areas ranging from 0.4 to 741,300 km². For avian and angiosperm species richness they used Wright's (1983) data, but compiled data on mammalian species richness themselves. Regression analysis established that for angiosperms, birds, and mammals, island energy explains more variation in species richness than area alone. However, the increase was not so great for mammals (4 per cent) as it was for angiosperms (55 per cent) and birds (56 per cent) in Wright's (1983) data set, probably because of the smaller island-size variation in the latter. When island area and island energy were introduced separately in multiple regression equations, they were both significant in explaining species richness of all mammals, mammals weighing less than 5 kg, and herbivorous species weighing more than 5 kg. No measure of island energy explained species richness of carnivores weighing more than 5 kg, possibly owing to the smaller sample size of carnivores. Applying the same multiple regression approach to Wright's data on birds and angio-sperms led to a greater explanation of the variance than did simple regression involving island energy alone. For all three groups of organism, the regression coefficients indicate that richness increases at a faster rate with increasing island energy than it does with island area (Table 8.2). Wright's model underestimates species richness at low latitudes and overestimates it at high latitudes, which might suggest that species richness varies with some variable other than energy that covaries with latitude. To examine this problem in more detail, Wylie and Currie entered island energy, island area, and latitude separately in a multiple regression equation. The results showed that, for angiosperms and birds, latitude does not explain a significant amount of species richness. In the case of mammals, the situation is not so clear. For instance, area, energy per unit area, and latitude all explain significant variation in the herbivore species richness, while area is the only variable explaining significant variation in carnivore species richness.

Island biogeography

When islands are colonized, the colonists seem to replace species that become extinct; in other words, there is a turnover of species. The theory of island biogeography is based upon the area effect, the distance effect, and species turnover. It thus combines geographical influences on species diversity with species change, and stresses the dynamism of insular communities. It was proposed independently by Frank W. Preston (1962), and Robert H. MacArthur and Edward O. Wilson (1963, 1967). Preston stressed the idea that island species exist in some kind of equilibrium. MacArthur and Wilson explicitly promulgated an equilibrium model. Their central idea is that an equilibrium number of species (animals or plants) on an island is the outcome of a balance between immigration of new species not already on the island (from the nearest continent) and extinction of species on the island. In other words, it reflects the interplay of species inputs (from colonization) and species outputs (from extinctions). The equilibrium is dynamic, the con-sequence of a constant turnover of species. In its simplest form, the MacArthur–Wilson hypothesis assumes that, as the number of species on an island mounts, so the rate of species immigration drops (because on average the more rapidly dispersing species would become established first, causing an initial rapid drop in the overall immigration rate, while the later arrival of slow colonizers would drop the overall rate to an ever-diminishing degree), and the rate of extinction of species rises (because the more species that are present, the more that are likely to go extinct in a unit time). The point at which the lines for immigration rate and extinction rate cross defines the equilibrium number of species for a given island (Figure 8.7a). To refine the model a little, MacArthur and Wilson assumed that immigration rates decrease with increasing distance from source areas: immigration occurs at a

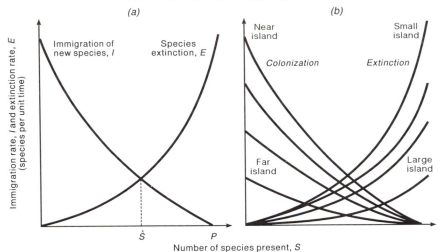

Figure 8.7 Basic relationships in the MacArthur–Wilson theory of island biogeography. (a) Equilibrium model of a biota of a single islands. The equilibrial number of species is defined by the intersection of the curves for the rate of immigration of species not already resident on the island, *I*, and the rate of extinction of species from the island, *E*. (b) Equilibrium models of biotas of several islands lying at various distances from the principal source areas of species and of varying size. An increase in distance (near to far) is assumed to lower the immigration rate, while an increase in area (small to large) is assumed to lower the extinction rate
Source: After MacArthur and Wilson (1963)

higher rate on near islands than on far islands (Figure 8.7b). For this reason, and all other factors being constant, the equilibrium number of species on a near island will be higher than that on a far island. In addition, MacArthur and Wilson assumed that extinction rates vary inversely with island size: extinction rates on small islands will be greater than extinction rates on large islands (Figure 8.7b). For this reason, and all other factors being constant, the equilibrium number of species on a small island will be lower than that on a large island. Several other refinements to the model were discussed in the monograph *The Theory of Island Biogeography* (MacArthur and Wilson 1967). These are summarized in Figure 8.8.

Several objections were raised to the original theory of island bio-geography (for reviews see Williamson 1981: 82–91; J. H. Brown 1986; Shafer 1990: 15–18). Some claimed that it is so oversimplified as to be useless (e.g. Sauer 1969; Lack 1970; Gilbert 1980). This is the normal and, to some degree, understandable response of many field-based researchers to mathematical models which reduce the complex patterns and processes encountered in the field to a few 'simple' formulae. Admittedly, a fuller description of change in island species richness, *S*, as a dynamic system could be written:

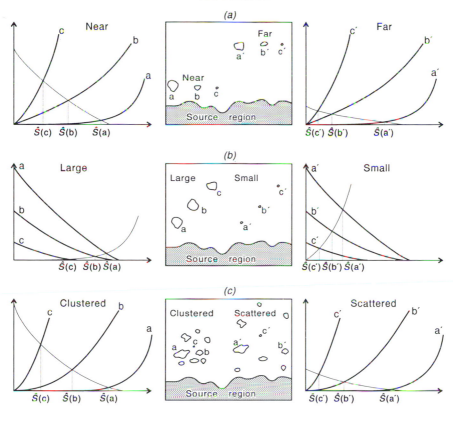

Figure 8.8 Refinements of the basic MacArthur–Wilson model of island biogeography

Source: After MacArthur and Wilson (1967)

$$\frac{dS}{dt} = (\text{Speciation} + \text{Immigration}) - (\text{Extinction} + \text{Emigration})$$

This equation is the equivalent to a population model in which population size changes on account of births and deaths, immigration and emigration. A potentially serious objection to MacArthur and Wilson's formulation was that it did not allow the possibility of autochthonous speciation – that is, the evolution of species on the island itself. Rather, it assumed, perhaps erroneously, that speciation is so slow compared with immigration that it can be ignored. This idea is testable. Within the Philippines, Lawrence R. Heaney (1986) found that speciation on islands has contributed substantially to species richness, exceeding colonization by a factor of two or more as a supplier of new species. So, autochthonous speciation is important on some islands.

MacArthur and Wilson assumed that immigration and extinction are the dominant processes on islands, and defined these as functions of distance from the source of new species and island area, respectively. It has been countered that immigration rate depends on island area as well as distance from mainland, since migrating animals are more likely to encounter a large island than a small one. Similarly, extinction rate may depend on distance from mainland as well as island area, since a dwindling species' population may be saved from extinction by the arrival of immigrants. This 'rescue effect' is likely to be more efficacious on near islands than far islands, and, where dominant, can lead to a higher turnover rate of species on far islands than on near islands, a phenomenon often observed in Nature (J. H. Brown and Kodric-Brown 1977). MacArthur and Wilson's curves for immigration and extinction rates are monotonic. It has been objected that, while immigration of animals species may decline monotonically with species richness, this is not the case with plant species where the rate of immigration is likely to be low for pioneer species and only rise to high levels once pioneers have established themselves. But such refinements are easily made to the basic model. Indeed, Michael E. Gilpin and Diamond (1976) devised and tested thirteen versions of the equilibrium model, each with a different combination of definitions for the rates of immigration and extinction. A more potent criticism of the MacArthur and Wilson model is that species numbers will be affected by habitat diversity, a factor that is not included in the model. It does seem fair to suggest that the resources of an island will support a limited number of species; in other words, there is a species carrying capacity that will varying according to environmental factors and the heterogeneity of an island landscape. Indeed, as mentioned earlier, the amount of energy received by an island is a good predictor of species richness (Wylie and Currie 1993a). Moreover, Wylie and Currie's island-based model accurately predicted the number of mammalian species confined to isolated nature reserves in the Australian wheatlands (Wylie and Currie 1993b).

In the MacArthur and Wilson model, immigration becomes zero when the island houses the same number of species that are stored in the mainland pool, but it seems unusual for an island to accommodate as many species as a continental region of the same size: islands are characteristically impoverished in species (e.g. Williamson 1981: 38). Actually, the model offers an explanation for the impoverishment of distant islands (compare near and far curves for a constant extinction rate in Figure 8.7b); but it is also possible that distant islands are impoverished in species because they are colonized more slowly and they have not yet been filled to a steady-state level. Another feature of island life is the mix of species, which is normally different from the mix of species on neighbouring mainland, partly because island biotas tend to be impoverished. The MacArthur–Wilson model is uninformative about the mix of species on an island. The poverty of island species becomes more pronounced towards the top of food chains. Differential impoverishment, combined with the fact that species have different dispersal abilities, leads to island biotas being disharmonious. Thus, oceanic islands tend to be populated by waif biotas and are characterized by disharmony: the Pacific island faunas contain hardly any mammals (except bats and rats); land birds are common, but most of them have a very restricted range, and only a few are migratory; reptiles are mainly represented by geckos and skinks which probably arrived on flotsam, while snakes are found only on islands near continents; amphibians and freshwater fishes are conspicuous by their absence; land molluscs are well represented, and the dominant animals are insects. Finally, it is reasonable to object that the equilibrium number of species will vary with time in response to environmental changes. Indeed, insular faunas may not normally be in equilibrium because geological and climatic changes can occur as fast as colonization and speciation (Heaney 1986). In other words, the basic model would usefully be fitted within the framework of a more inclusive set of equations, such as the 'brash' equation.

Despite these criticisms, many of which can be overcome by minor modifications of the basic model, the theory of island biogeography has engendered much valuable debate and a flood of field investigations into the effect of insularity within geoecosystems. Indeed, many studies tend to vindicate the basic thesis that steady-state species numbers are a function of area and distance. Kenneth L. Crowell (1986), for instance, found that mammalian faunas on twenty-four land-bridge islands in the Gulf of Maine could be explained by recurrent colonizations, mainly via ice bridges. Species richness varied directly with area and inversely with distance, though the richness was modulated by climatic change, range expansions, and human disturbance. Other studies of mammalian faunas on islands in Finnish lakes (Hanski 1986) and in the St Lawrence River (Lomolino 1986) show that they are roughly in equilibrium with colonization rates balancing extinction rates.

The theory of island biogeography has indisputably stimulated new ideas about species in habitat islands. An interesting idea is that the structure of insular mammalian faunas in a state of 'relaxation' may be described and explained by the nested subset hypothesis. First proposed by Bruce D. Patterson (1984), this hypothesis holds that the species in a depauperate insular fauna should consist of a proper subset of those in richer faunas, and that an archipelago of such faunas arranged by species richness should present a nested series (Patterson 1984, 1987, 1991; Patterson and Atmar 1986; Patterson and Brown 1991). A case in point is the debate over small mammal populations in montane islands in the American Southwest. It was originally thought that these montane islands contain 'relaxation' faunas in which species richness decreased ('relaxed') through time as species demanding large resources (such as big specialist carnivores) became extinct and, owing to isolation, no immigrant species made good the losses (J. H. Brown 1971). The argument ran that the montane islands of forests and woodlands are fragments of a once continuous forest–woodland habitat that connected boreal habitats of the Sierra Nevada, Great Basin, and Rocky Mountains. After the Pleistocene epoch ended, climatic warming and drying broke up the forest–woodland which eventually contracted to higher elevations where relatively cool and moist local climates could support forest and woodland. Small, non-volant mammals living in these montane forest islands should thus be influenced by selective extinctions during relaxation, but not by Holocene immigrations of new species. The fact that mammalian species richness of these habitat islands in the Great Basin was correlated with island area (a measure of extinction probability), but not isolation (a measure of immigration potential), lent support to this hypothesis. A new hypothesis for the structure of mammalian communities on montane islands, that includes the possibility of Holocene immigration, was proposed when it was recognized that the great majority of montane islands in the American Southwest are now isolated by woodland and not grassland, chaparral, or desert scrub (Lomolino *et al.* 1989). Strong support for the hypothesis came from highly significant correlations between mammalian species richness and current isolation. Furthermore, when isolation was partitioned between distance to be travelled across grassland and chaparral habitats versus distance to be travelled across woodland, species richness was significantly correlated with the former but not with the latter, showing that woodlands do not present a major obstacle to immigration. Undoubtedly, historical

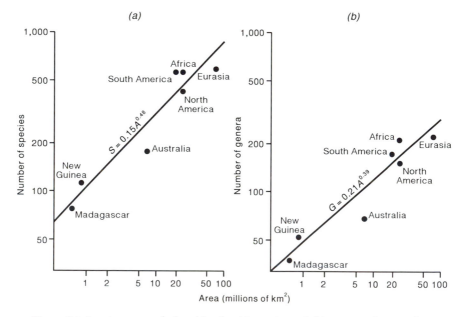

Figure 8.9 Species–area relationships for (a) species and (b) genera of non-volant terrestrial mammals inhabiting continents and large islands
Source: After J. H. Brown (1986)

222

processes of vicariance and subsequent selective extinction have helped to structure these communities, but the combined importance of immigration and extinction argues for a dynamically structured system on the lines set out by MacArthur and Wilson (1967).

The conjoint action of immigration and extinction may be important in other systems formed by vicariance events, such as land-bridge islands created by rising sea-level, if isolation is not great enough to create a true barrier to dispersal. There is evidence that entire continents may be regarded as islands subject to processes of colonization, speciation, and extinction. As James H. Brown (1986: 241) remarked, 'from a biogeographical perspective continents are nothing more than very large islands with complex histories of climatic and geological change'. Karl Flessa (1975, 1981) showed that continental patterns of the richness and affinities of genera indicate underlying mechanisms of colonization and extinction similar to those predicted by the MacArthur–Wilson model, the relation between number of genera and area being described by a power curve and suggesting that increasing extinction rates play a substantial role in limiting diversity on progressively smaller land masses. Confirmation of these results came from Brown (1986) (Figure 8.9). So, the basic model of island biogeography, although it needs elaborating, can be applied to all scales of geoecosystems, from mites on mice to mammals on continents.

SUMMARY

The relative isolation of a geoecosystem – its insularity – is a topospheric feature that has a significant influence on individuals and communities. True islands have abnormally high rates of extinction, house many endemic and relict species, favour the evolution of dwarf, giant, and flightless forms, and are often inhabited by good dispersers. Island communities normally have steeper species–area curves than do mainland communities. Insular species–area relationships may simply result from the effect of increasing area, but increasing habitat diversity with increasing area appears to play a role. Species richness on islands has also been shown to depend on energy receipt as measured by solar radiation, potential evapotranspiration, actual evapotranspiration, and net primary productivity. The theory of island biogeography proposes that species turnover on islands depends on colonization rates and extinction rates, both of which are assumed to vary with island area and distance from the mainland. Although severely criticized in some quarters, the theory of island biogeography has provoked much valuable discussion and prompted a great many investigations. It has been found to be applicable in many cases and across the full range of spatial scales, from microscale to megascale.

FURTHER READING

Patrick D. Nunn's *Oceanic Islands* (1993) is an commendable introduction to the physical geography of islands. The best general introduction to life on islands is still Carlquist's *Island Life: A Natural History of the Islands of the World* (1965). MacArthur and Wilson's monograph, *The Theory of Island Biogeography* (1967), remains a landmark in the evolution of ideas on the effects of insularity. Read it! A plain, simple, and readable account of life on islands is found in Martyn L. Gorman's *Island Ecology* (1979), while Mark Williamson offers somewhat meatier fare in his *Island Populations* (1981). Volume 28 of the *Biological Journal of the Linnean Society* deals with mammalian island biogeography and should not be missed. A very useful source of ideas about, and references on, island biogeography (both true islands and habitat islands) is Craig L. Shafer's *Nature Reserves: Island Theory and Conservation Practice* (1990).

Part III

EXTERNAL INFLUENCES

9

DISTURBANCE

Disturbance has been a popular theme in ecological studies over the last two decades (e.g. Barrett *et al.* 1976). A working definition of disturbance is any event that disrupts the ecosphere or an ecospheric system. In the present context, it is any event that disrupts a geoecosystem. It would include environmental fluctuations and destructive events (White and Pickett 1985: 6). In the 'brash' formula, disturbance is a driving variable, even though it may originate from within the geoecosphere. Furthermore, disturbing agencies may be physical or biological. Grazing, for example, may be defined as a biotic driving variable of vegetation communities, although it belongs to the same geoecosystem as the plants that it disturbs. Plainly, disturbance may result from external or internal geoecosystem processes. It is useful, however, to think of disturbance as a continuum between purely endogenous processes and purely exogenous processes (White and Pickett 1985: 8). At the exogenous end of the continuum is classical 'disturbance'. This is demonstrably the outcome of exogenous processes acting at a particular time, creating sharp patch boundaries, and increasing resource availability. At the endogenous end of the continuum is internal dynamics of communities, exemplified by Watt's (1947) cycles of vegetation change.

Landscapes may be physically disturbed by strong winds (which uproot trees), fire, flood, landslides, lightning, and the impact of extraterrestrial bodies; and biotically disturbed by pests, pathogens, and the activities of animals and plants, including the human species. The effects of these disturbing agencies on ecosystems can be dramatic. Pathogens, for example, are forceful disrupters of ecosystems at small and medium scales, witness the efficacy of Dutch elm disease in England and the chestnut blight in the Appalachian region, eastern United States. Chestnut blight is caused by a fungal parasite, *Endothia parasitica*. Introduced into New York on imported Asiatic chestnut stock, the chestnut blight practically destroyed all the chestnut timber in the Appalachian region within thirty years of its discovery in 1904. This was an ecological disturbance of catastrophic proportions: the most important canopy species in the oak–chestnut forest was removed. Less potent disturbance may still lead to significant changes in species and communities.

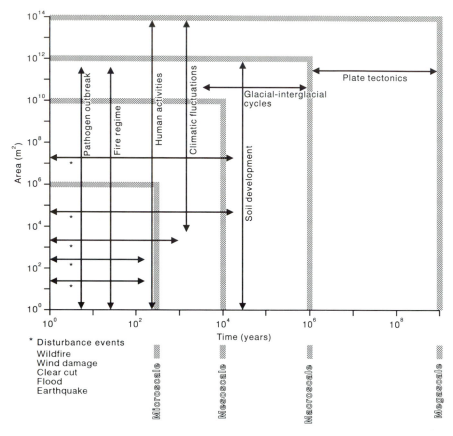

Figure 9.1 Space and time domains of landscape systems and disturbing agents
Source: After Delcourt and Delcourt (1988)

Landscapes at all scales, both spatial and temporal, are subject to disturbance. Paul A. and Hazel R. Delcourt (1988) diagnosed the chief disturbing agents at four different scales (Figure 9.1). At microscales, defined as 1 to 500 years and covering a square metre to a square kilometre, wildfire, wind damage, clear cut, flood, and earthquake are the dominant causes of disturbance events. Vegetational units at this scale range from individual plants and forest stands, and landscapes range from sample plots to first-order drainage basins. Local disturbances lead to patch dynamics within individual vegetation patches. Disturbance events over mesoscales, defined as half a millennium to ten millennia and a square kilometre to ten thousand square kilometres, encompass interglacial stages and landscapes ranging from second-order drainage basins to mountain ranges up to 1° × 1° (latitude × longitude). On the lower end of this scale, a prevailing disturbance regime, such as pathogen outbreaks and frequent fires, influences patch dynamics over a landscape mosaic. In the upper range of the scale, the prevailing disturbance regimes may themselves change so causing changes within patches and between patches that in turn alter the landscape mosaic. Disturbances at macroscales, defined as ten millennia to a million years, and ten thousand square kilometres to a million square kilometres, span one to several glacial–interglacial cycles and affect landscapes ranging in size from physiographic provinces to subcontinents. At this scale, changes in prevailing disturbance regime are effected by regional and global environmental changes. Disturbances at the largest scale, defined as a million years to 4.6 thousand million years, and areas more than a thousand square kilometres (continents, hemispheres, the globe), are driven by plate tectonics which alters global climates and influences biotic evolution. It would be useful to add bombardment and volcanic activity to this scheme. Bombardment disrupts ecosystems within seconds, disturbing small or very large areas depending on the size of the impactor. Astronomical theory suggests that the bombardment regime may be periodic. Volcanic activity, like bombardment, acts very quickly and may disturb the entire ecosphere, though local disturbances are far more common. Repeated disturbance during a bout of volcanism is, arguably, more likely to cause severe stress in the ecosphere, than is a single volcanic explosion.

ECOLOGICAL DISTURBANCE

The process of disturbance may be viewed in two ways (M. G. Turner and Dale 1991). First, some disturbances act essentially randomly within a landscape to produce disturbance patches. Strong winds commonly behave in this way. The patches produced by random disturbance can be extensive: the biomantles formed by burrowing animals are a case in point (D. L. Johnson 1989, 1990), as are the patches of eroded soil created by grizzly bears excavating dens, digging for food, and trampling well-established trails (D. R. Butler 1992). Second, some disturbances, such as fire, pests, and pathogens, tend to start at a point within a landscape and then spread to other parts. In both cases, the disturbances operate in a heterogeneous manner because some sites within landscapes will be more susceptible to disturbance agencies than others. A study of the old-growth Pisgah Forest, south-western New Hampshire, United States, revealed that disturbance in the period 1905 to 1985 depended on slope position and aspect (D. R. Foster 1988a), and wind damage in forest stands was related to the age of trees (D. R. Foster 1988b). In Yellowstone National Park, older stands of coniferous forest are more susceptible to fire than younger stands (Romme 1982). Another general point to make is that the agencies of disturbance often work in tandem (e.g. Veblen *et al.* 1994). In forested landscapes of the south-eastern United States, individual pine trees are disturbed by lightning strikes. Once struck, pine trees are susceptible to colonization by bark beetles whose populations can expand to epidemic proportions and create forest patches in which gap-phase succession is initiated; thus the bark beetle appears to magnify the original disturbance by lightning (Rykiel *et al.* 1988). To illustrate these points, disturbance by wind, fire, and grazing will be examined. Many other forms of disturbance could be discussed, but lack of space precludes comprehensive coverage. However, two studies of disturbance regimes will be mentioned as they show how multivariate analysis and dynamic systems models are being used to study the effect of environmental change on geoecosystems.

Wind

Tree-throw by strong winds (Plate 9.1), and to lesser extent other factors, can have a considerable impact on landscapes. In Puerto Rico, hurricanes appear to maintain the composition and structure of lower-montane rain forest (Doyle 1981). Indeed, in many forests, disturbance by uprooting is the primary means by which species richness is maintained (Schaetzl *et al.* 1989a, 1989b). A fallen tree creates a gap in the forest that seems vital to community and vegetation dynamics and successional pathways: it provides niches with

Plate 9.1 A wind-thrown tree on the Kingston Lacey estate, Dorset, England.
Photograph by Shelley S. Huggett

Plate 9.2 A gap in a forest, Taiwan, caused by a tropical cyclone.
Photograph by Ian Douglas

much sunlight for pioneer species, encourages the release of suppressed, shade-tolerant saplings, and aids the recruitment of new individuals. In 1938, a hurricane travelled across New England and destroyed whole stands of trees. Some of the trees had their trunks broken, but most were uprooted. On the ground, the fallen trees created pit-and-mound topography that influenced the distribution of understorey plants through microclimatological and microtopographic effects. The distribution of tree seedlings was related to microtopography through soil morphology, nutrition, and moisture content at pit-and-mound sites. In Harvard Forest, Petersham, Massachusetts, sixty-two pit-and-mound pairs were found (E. P. Stephens 1956). Investigation revealed that they were formed in four episodes: 1938, 1815 (during a hurricane on 20 September that year), sometime in the first half of the seventeenth century (possibly 1635 when a hurricane was recorded in the Plymouth Colony), and the second half of the fifteenth century. Even in areas with no pit-and-mound microtopography, soil horizons invariably showed traces of overturned horizons, a sure sign of much older wind throws. In the forests of eastern North America generally, disturbance has caused curious stands of trees, all individuals in which are the same age, in remnants of presettlement forests. These have a patchy distribution owing to 'the varying paths of storms and to the fickle behaviour of the winds in local areas' (H. M. Raup 1981: 43).

Plate 9.3 Burning scrub, Manukaia Waikato, New Zealand.
Photograph by Brain S. Kear

Tropical forests, once thought immune to physical disturbance, have been shown to contain seemingly haphazard patterns of tree forms, age classes, and species. Timothy C. Whitmore (1975) suggested that the tropical rain forests of the Far East should be analysed as 'gap phases'. Gaps are openings made in the forests by various disturbances, and the phases are the stages of tree growth in the gaps, from seedling to maturity and death (Plate 9.2). Small gaps, about 0.04 ha in area, were opened up by the fall of individual large trees, with crowns 15 to 18 m in diameter. Lightning strikes opened up gaps with an area of about 0.6 ha. Local storms caused larger gaps (up to 80 ha), while typhoons and tornadoes destroyed even larger areas. These gaps are an integral part of the tropical forest system. Later work has shown that community and population dynamics in tropical forests could only be explained if disturbance was included (e.g. Hubbell 1979), and that mosaic cycles, often driven by disturbance, are common in many communities (e.g. Brokaw 1985; Remmert 1991).

Fire

Fire is relatively common in many terrestrial environments (Plate 9.3). It has long been recognized that fire influences the structure and function of some plant species and many communities (e.g. Ahlgren and Ahlgren 1960; C. F. Cooper 1960; Kozlowski and Ahlgren 1974). In some cases, the effects of fire appear to outweigh climatic effects. The elevational distribution of three pine species (*Pinus discolor, Pinus leiophylla,* and *Pinus engelmannii*) in the Chiricahua Mountains, Arizona, is not, as would be expected, determined by light levels, shade tolerance, and drought resistance, but chiefly by fire (Barton 1993). Similarly, fire greatly influences the structure and composition of certain tree stands along an elevational gradient in the central Himalayas (Tewari *et al.* 1989).

A fine example of experimental work that investigates the effects of fire disturbance is the research carried out under the auspices of the Fynbos Biome Project in the Swartboskloof catchment, near Stellenbosch, Cape Province, South Africa (van Wilgen *et al.* 1992). The vegetation in the 373 ha catchment is dominantly mesic mountain fynbos, a Mediterranean-type community. Forests grow in wetter sites around perennial streams and on boulder screes below cliffs. Major fires occurred in 1927, 1942, and 1958. These fires burnt the entire catchment. Smaller fires burnt parts of the catchment in 1936, 1973, and 1977. All the fires were accidental, save the prescribed burn in 1977. The research project studied the effects of fires on single species, on communities, and on the entire catchment geoecosystem (D. M. Richardson and van Wilgen 1992). At the population level, it was found that mountain fynbos plants possess a wide range of regeneration strategies and fire-survival mechanisms. Most species can sprout again after a fire, and are resilient to a range of fire regimes. Few species are reliant solely on seeds for regeneration. Those that are obligate reseeders regenerate from seeds stored in the soil, and just a few, such as *Protea neriifolia,* maintain seed stores in the canopy. Most of the fynbos plants flower within a year of a fire. Some species, including *Watsonia borbonica* and *Cyrtanthus ventricosus,* are stimulated into flowering by fires, though the trigger appears to act indirectly through changes in the environment (such as altered soil temperature regimes), and not directly through heat damage to leaves or apical buds. Different groups of plants in the mountain fynbos respond differently to changes in fire regimes. For example, fire intensity affects the balance between sprouters, which are more sensitive to high intensity fires, and reseeders. At the community level, there was no evidence that fire played a role in shaping the boundaries within the fynbos vegetation groups, but it did exert a major influence on the boundary between fynbos and forest. Interestingly, it was found that fire led to an increase in community invasibility; that is, it left communities more open to invasion by indigenous and alien taxa. The conditions under which invasion occurred were determined by the frequency, season, and intensity of fires. Variability of the fire regime has allowed a host of species to coexist, but when the regimes are disrupted, either by excluding fire or by introducing alien species, new stable communities may evolve. At the geoecosystem level, studies in Swartboskloof considered the effects of fires on nutrient budgets, streamflow, and erosion. Results showed that fires led to increased overland flow and increased sediment yield in small plots, but the increases were of short duration and did not lead to changes in streamflow or sediment yield in the entire catchment. This lack of response at the geoecosystem level was ascribed to the riparian forests, which act as a buffer against overland flow and erosional loss. Overall, the studies in the Swartboskloof catchment illustrate the value of investigating geoecosystems and their components at a variety of scales and levels of organization.

Fires in natural vegetation are undoubtedly influenced by climatic factors. Warmer and drier climates tend to favour forest fires, though the link between climate and wildfires is uncertain (Christensen *et al.* 1989; Romme and Despain 1989; Swetnam and Betancourt 1990). There is a suggestion that wildfires will increase if the globe warms up during the next century. Analysis of historical records of fires and climate in Yellowstone National Park, United States, from 1895–1989, reveals that 36 per cent of the temporal variance in annual burn area is explained by surface climate variables that induce droughtiness: increasing temperatures in the fire season; decreasing precipitation in the antecedent season; and, a trend to drought conditions in the antecedent season (Balling *et al.* 1992). The wildfires disturb biological systems and greatly enhance the sediment transport through the landscape. Analogues of the debris flow deposits and sedimentation events in streams following the 1988 fires in the park (Christensen *et al.* 1989) have been discovered in older alluvial fan deposits in the north-eastern section, allowing past sedimentation events related to fires to be studied (Meyer *et al.* 1992). It was found that alluvial fans aggrade during frequent sedimentation events related to fires, and that periods of fire outbreaks appear to be associated with times of drought or high climatic variability. During wetter periods, sediment is removed from alluvial fan storage and transported down streams, resulting in floodplain aggradation. In short, the dominant fluvial activity in the area is modulated by climate, with fire, through its disturbing action on vegetation, acting as a catalyst for sediment transport during droughts.

Over the last century, there has been a trend to a set of climatic conditions that favour the outbreak of fires in Yellowstone National Park (Balling *et al.* 1992). Whether this is a greenhouse signal is not clear. Numerical climate models predict increasing aridity in the area associated with higher rainfall but even higher evapotranspiration. The historical records for the area show that aridity appears to relate to increased temperature, which is generally consistent with climate model predictions, and a decrease in precipitation in the antecedent season, which is not consistent with the climate model predictions. Whatever its cause, the trend towards increasing aridity enhances the likelihood of fires, as is clear from the historical records.

Mathematical models have helped in understanding the nature of fires and their impact. A model of the daily spread of a fire front in Glacier National Park used six environmental gradients (elevation, topographic moisture, time since last burn, primary succession, drainage, and alpine wind-snow exposure), and four categories of disturbance (intensity of last burn, slide disturbances, hydric disturbances, and influences of heavy winter grazing by ungulates) (Kessell 1976, 1979). The procedure was to use gradient analysis to estimate the vegetation and fuels present throughout a large area, then run the fire model in each spatial cell, and then simulate post-fire succession deterministically according to habitat types (Kessell 1979) or to life history traits of the species (Cattelino *et al.* 1979). Other models consider the spatial patterns of fire and species dispersal, but do not incorporate the actual spread of fire. An example is a model of forest fires which used a 50×50 square grid, each grid-cell representing areas with diameters of 10 to 100 m, depending on the disturbance rules adopted (Green 1989). Each grid-cell was assumed to contain a single plant (at the smallest spatial scale) or a stand (at the larger spatial scales). Fires were ignited at random locations and were elliptical in shape, their frequency being determined by sampling from a Poisson distribution, and their size being determined by a negative exponential distribution. Simulated fires created non-uniform patches into which species could disperse, so building up a mosaic of different vegetation types that varied according to whether dispersal was influenced by an environmental gradient. Another model – the BRIND model – combined a forest growth model with several phenomena, including fire, that are important in Australia (Shugart and Noble 1981). The model was built to explore the long-term dynamics of *Eucalyptus*-dominated forests in the Brindabella Range, near Canberra, Australia. Altitudinal differences in community composition were predicted: at 850 m, mixed stands of *Eucalyptus robertsonii*, *E. rubida*, and *E. viminalis* occurred; at 1,050 m, *E. robertsonii* and *E. fastigata* became dominant; at 1,300 m, pure stands of *E. delegatensis* or, in some simulation runs, stands dominated by either *E. dalrympleana* and *E. pauciflora*, arose; and at 1,500 m, on mountain peaks, stands dominated by *E. pauciflora* appeared, either as pure stands or in mixtures with *Bouksia marginata* and *Acacia dealbata*. These different communities resulted from altitudinal variations in degree-days (ranging from about 2,400 at 850 m to 1,200 at 1,500 m) and from wildfire probabilities. They closely matched observed tree communities in the Brindabella Range.

Grazing

At least since the time of Charles Darwin, it has been known that grazing by herbivores alters the structure and composition of plant communities. Controversy surrounds the role of herbivory in ecosystems. Proponents of the food-limitation hypothesis claim that herbivorous animals regulate the abundance of plant populations (Brues 1946) and may control the functioning of entire ecosystems (Chew 1974), and that the abundance of most herbivore populations is limited by food. Opposing beliefs are that herbivores are normally scarce relative to their food supply owing to the

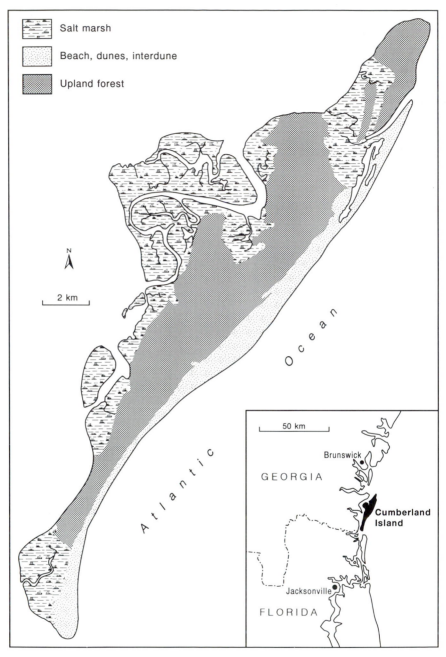

Figure 9.2 Vegetation on Cumberland Island, Georgia, USA
Source: After Turner and Stratton (1987)

depredations of natural enemies (Hairston *et al.* 1960) and the weather (Andrewartha and Birch 1954), or that herbivory is, in the main, beneficial to plants (Owen 1980). These issues are not settled, but grazing is without doubt an important element in nearly all geoecosystems, and may greatly influence community stability and productivity (see McNaughton 1993). The overall effects of grazing have been ascertained by compiling data from a range of environments (Milchunas and Lauenroth 1993). A world-wide set of data was compiled at 236 sites where species composition, aboveground net primary production, root biomass, and soil nutrients in grazed versus protected ungrazed plots had been measured. The data set was subjected to multivariate regression analysis. It was found that changes in species composition with grazing depended primarily on above-ground net primary production and the evolutionary history of the grazing site.

A complication in establishing the effects of grazing is that it often acts in concert with other disturbing agencies. This is true on Cumberland Island, a barrier island lying off the coast of Georgia, United States, where the propagation and behaviour of disturbances caused by grazing and fire was studied by Monica Goigel Turner and Susan P. Stratton (1987). Cumberland Island is composed of a core of stable Pleistocene sediments, rarely subject to storm overwash, surrounded by a periphery of more dynamic Holocene beach, dune, and marsh sediments that are readily refashioned by storm events. Vegetation on the island is arranged in longitudinal belts that are chiefly the result of physical influences such as tidal energy, salt spray, and topography (Figure 9.2). The upland forests tended to be patchy. During the present century, horses (*Equus caballus*), cattle (*Bos bovine*), and hogs (*Sus scrofa*) roamed free on the island, until in 1974 the cattle were removed and the hog population reduced through trapping. Some 180 horses still live there, and the native white-tailed deer (*Odocoileus virginianus*) population has grown large following release from hunting and removal of predators. Intensive grazing by these animals has depleted the forest understorey and clipped interdune and high marsh vegetation to the height of a mown lawn (Plates 9.4a, b). Besides grazing disturbance, large fires on the island have been sparked by lightning, especially during dry summers. Approximately 1,300 hectares have been burnt since 1900, much of it repeatedly on a 20- to 30-year rotation. Fire size varies with community type: in scrub, fires are large, hot, and often burn into other vegetation types; in mature pine stands and freshwater marshes, fires are less intense, although they may still burn into fringing areas.

Taken together, fire and grazing are important determinants of the landscape mosaic on Cumberland Island, and, although they behave very differently, they both 'consume' biomass as they proceed and, owing to the influence of landscape heterogeneity, induce a similar response in the landscape. Feral horses mainly graze the salt marshes and grassland, which are both highly disturbed and resilient, but also influence the maritime forest

Plate 9.4a Effects of grazing on Cumberland Island, Georgia: a live oak (*Quercus virginiana*) forest with virtually no understorey.
Photograph by Monica G. Turner

Plate 9.4b Effects of grazing on Cumberland Island, Georgia: an extensive area of grazed marsh.
Photograph by Monica G. Turner

Plate 9.5 Horses grazing an interdune area on Cumberland Island.
Photograph by Monica G. Turner

which cannot by itself support the horse herd (Plate 9.5). The salt marshes and grassland are energy sources for a perturbation that may change the forest community. Deer graze the interdune meadows, open grassy fields, and upper edges of the salt marshes, and will browse most of the broad-leaved woody species on the island. Disturbance arises when the deer population grows and it moves from its preferred habitats to other landscape patches. As browsing by deer is selective, it may change forest composition. Fires will only start in communities with adequate fuel supplies. On Cumberland Island they generally begin in the scrub or freshwater marsh where most of the available plant material is consumed. The scrub, and the freshwater marshes, are highly resilient and replace themselves quickly after a burn. The oak forests are fire resistant, but when burned recover slowly. Scrub patches within the landscape create a focus for fire that can move into the forest and convert it to scrub. In conclusion, fire and grazing disturbances on Cumberland Island seem to be driven by resilient patches within the landscape, yet influence resistant patches from which little energy is obtained. It should be pointed out that fire and grazing are only part of the disturbance regime on the island – feral pigs, human visitors, and storms also play a part.

The effects of grazing on geoecosystems are complex. This fact is underscored by dynamic systems models of plant–herbivore dynamics that use a variety of equation akin to the 'brash' equation. An example is a model of plant–herbivore interactions in which plant growth is constrained by a carrying capacity (Crawley 1983: 249):

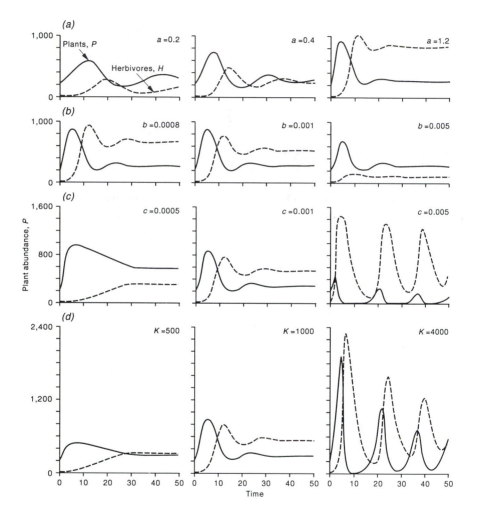

Figure 9.3 Dynamics of a plant–herbivore system with resource-limited plants (growth constrained by an environmental carrying capacity). (a) The effect of increasing the growth rate of plants, a ($b = 0.001$, $c = 0.001$, $d = 0.3$, $K = 1,000$). (b) The effect of increasing herbivore searching efficiency, b ($a = 0.5$, $c = 0.001$, $d = 0.3$, $K = 1,000$). (c) The effect of increasing the carrying capacity, K ($a = 0.5$, $b = 0.001$, $d = 0.3$, $K = 1,000$). (d) The effect of increasing herbivore searching efficiency, b ($a = 0.5$, $b = 0.001$, $c = 0.001$, $d = 0.3$)
Source: After Crawley (1983)

240

$$\frac{dP}{dt} = \frac{aP(K - P)}{K} - bHP$$

$$\frac{dH}{dt} = cHP - dH$$

In these equations, P is the plant population, H is the herbivore population, K is the carrying capacity of plants, and a, b, c, and d are parameters: a is the intrinsic growth rate of plants in the absence of herbivores; b is the depression of the plant population per encounter with a herbivore, a measure of herbivore searching efficiency; c is the increase in the herbivore population per encounter with plants, a measure of herbivore growth efficiency; and d is the decline in herbivores (death rate) in the absence of plants. The parameters and the carrying capacity affect system stability and steady-state populations numbers. In brief, increasing the intrinsic growth rate of plants, a, increases system stability and herbivore steady-state density, but has no effect on plant abundance (Figure 9.3a). Increasing herbivore searching efficiency, b, reduces, somewhat surprisingly, herbivore steady-state density, but has no effect on system stability or on steady-state plant abundance (Figure 9.3b). Increasing herbivore growth efficiency, c, reduces steady-state plant abundance and so reduces system stability, but increases steady-state herbivore numbers (Figure 9.3c). Increasing the carrying capacity of the environment for plants, K, increases herbivore steady-state density and reduces system stability, but, paradoxically, has no effect on steady-state plant abundance (Figure 9.3d).

Clearly, the dynamics displayed by the plant–herbivore model, even in such a relatively simple case, are complex. The system's dynamics are further complicated by adding more realistic features, such as more complex functional responses of herbivores, plant compensation, and impaired plant regrowth. Importantly, some features of the dynamics, such as the paradox of enrichment (whereby an increased plant-carrying capacity has no effect on steady-state plant abundance), are counterintuitive. This is why dynamical systems models make such a valuable accompaniment to field investigations.

Disturbance regimes

Disturbance is an integral part of geoecosystems. It seems sensible, therefore, to include it as a system variable. Robert J. Whittaker's (1987, 1989) study of vegetation–environment relationships on the Storbreen glacier foreland, Jotunheimen, Norway, did just that, and revealed the intricate connections within the geoecosystem complex. The foreland has been recently deglaciated and the ages of different parts of the land surface are known (Plate 9.6). Indirect and direct gradient analysis, as defined by Robert H. Whittaker

Table 9.1 Correlation matrix for vegetation and environmental variables on Storbreen glacier foreland, Jotunheimen, Norway

	Terrain age	Altitude	Maximum slope	Moisture regime	Overall disturbance	Exposure	Snowmelt	Frost churning	Slope movement	Litter depth	Soil depth	Root depth	DCA axis I
Altitude	-0.635[a]												
Maximum slope	0.157	-0.059											
Moisture regime	-0.132	0.023	-0.139										
Overall disturbance	-0.594	0.452	-0.066	0.152									
Exposure	0.177	-0.348	0.088	0.298	-0.175								
Snowmelt	-0.137	0.001	0.128	0.341	0.086	0.499							
Frost churning	-0.684	0.508	-0.122	0.058	0.776	-0.349	-0.027						
Slope movement	-0.193	0.149	0.361	-0.052	0.438	-0.061	0.176	0.359					
Litter depth	0.360	-0.230	0.206	-0.005	-0.378	0.142	0.072	-0.426	-0.080				
Soil depth	0.493	-0.335	0.211	-0.047	-0.497	0.159	0.036	-0.527	-0.122	0.806			
Root depth	0.279	-0.218	0.080	-0.021	-0.360	0.139	0.008	-0.301	-0.183	0.365	0.455		
DCA axis I	0.762	-0.566	0.134	-0.208	-0.590	0.131	-0.174	-0.659	-0.202	0.319	0.488	0.261	
DCA axis II	0.240	-0.210	0.142	0.311	-0.216	0.447	0.347	-0.417	-0.040	0.296	0.278	0.256	0.171

Note: [a]Kendall's τ, correlations significant at $p \leq 0.001$ are emboldened
Source: After R. J. Whittaker (1987)

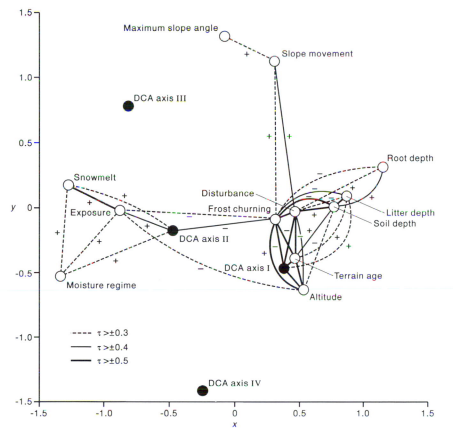

Figure 9.5 Plexus diagram defined by non-metric multidimensional scaling of vegetation and environmental variables, Storbreen glacier foreland, Jotunheimen, Norway

Source: After R. J. Whittaker (1987)

variables and the four vegetation gradients were computed as rank correlation coefficients (Table 9.1). The use of a non-parametric correlation coefficient was justified because some of the variables were measured on an ordinal scale, and it was possible that non-linear relationships between variables were present in the data set. A plexus diagram was then constructed from the matrix of correlation coefficients using non-metric multidimensional scaling (Figure 9.5). Two of the four vegetation axes closely relate to two environmental factor complexes. The 'terrain-age factor complex' (comprising terrain age, frost churning, disturbance, altitude, and the three soil variables) is closely associated with DCA axis I, the most important vegetation axis ($\lambda_1 = 0.747$). The 'exposure–moisture–snowmelt (micro-topographic) factor complex' is associated with DCA axis II ($\lambda_2 = 0.0.277$).

(a)

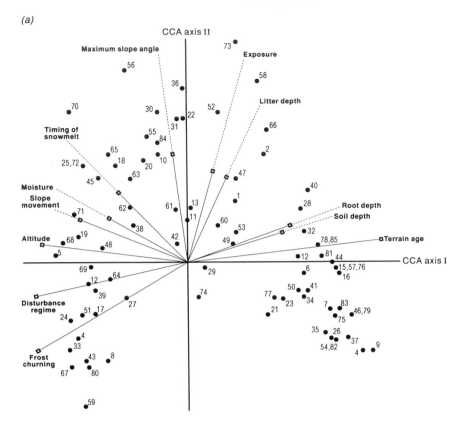

Figure 9.6 Biplots of CCA vegetation axes showing the position of
species and environmental variables, Storbreen glacier foreland, Jotunheimen,
Norway. The species, identified by numbers, are: 1 *Antennaria alpina.*
2 *Anthoxanthum odoratum.* 3 *Arabis alpina.* 4 *Arctostaphylos alpina.*
5 *Astragalus alpinus.* 6 *Bartsia alpina.* 7 *Betula nana.* 8 *Betula pubescens*
(including *B. tortuosa*). 9 *Campanula rotundifolia.* 10 *Cardamine bellidifolia.*
11 *Cardaminopsis petraea.* 12 *Carex* spp. 13 *Cassiope hypnoides.*
14 *Cerastium* spp. (*C. alpinum* and *C. cerastoides*). 15 *Coeloglossum viride.*
16 *Comarum palustre.* 17 *Cryptogamma crispa.* 18 *Cystopteris fragilis.*
19 *Deschampsia alpina.* 20 *Deschampsia cespitosa.* 21 *Deschampsia flexuosa.*
22 *Dryas octopetala.* 23 *Empetrum hermaphroditum.* 24 *Epilobium
alsinifolium.* 25 *Epilobium anagallidifolium.* 26 *Equisetum* spp. 27 *Erigeron
uniflorum.* 28 *Eriophorum angustifolium.* 29 *Festuca* spp. (*F. ovina*).
30 *Gnaphalium norvegicum.* 31 *Gnaphalium supinum.* 32 *Hierachium* spp.
33 *Juncus biglumis.* 34 *Juncus trifidus.* 35 *Juniperus communis.* 36 *Leontodon
autumnalis.* 37 *Loiseleuria procumbens.* 38 *Luzula* spp. (*L. arcuata,*

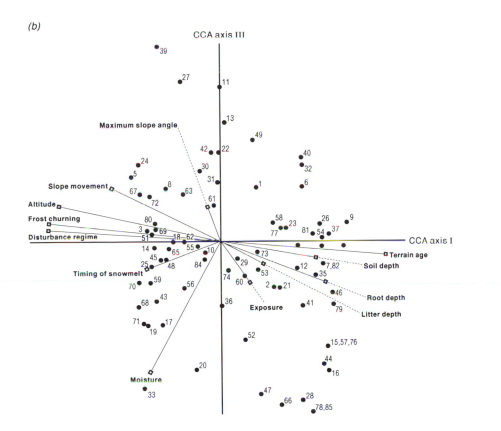

Figure 9.6 (continued) *L. confusa, L. frigida, L. spicata*). 39 *Lychnis alpina*
(*Viscaria alpina*). 40 *Lycopodium alpinum*. 41 *Lycopodium annotinum*.
42 *Lycopodium selago*. 43 *Melandrium apetalum*. 44 *Nardus stricta*.
45 *Oxyria digyna*. 46 *Pedicularis lapponica*. 47 *Pedicularis oederi*. 48 *Phleum*
commutatum (*P. alpinum*). 49 *Phyllodoce caerulea*. 50 *Pinguicula vulgaris*.
51 *Poa alpina* subspecies *alpina* and *vivipara*. 52 *Polygonum viviparum*.
53 *Pyrola* spp. (*P. minor, P. norvegica, P. rotundifolia*, and *Orthilia secunda*).
54 *Ranunculus acris*. 55 *Ranunculus glacialis*. 56 *Ranunculus pygmaeus*.
57 *Rumex acetosella*. 58 *Rumex acetosa*. 59 *Sagina saginoides*. 60 *Salix*
glauca. 61 *Salix herbacea*. 62 *Salix lanata*. 63 *Salix myrsinites*. 64 *Salix*
phylicifolia. 65 *Salix reticulata*. 66 *Saussurea alpina*. 67 *Saxifraga cespitosa* (*S.*
groenlandica). 68 *Saxifraga nivalis*. 69 *Saxifraga oppositifolia*. 70 *Saxifraga*
rivularis. 71 *Saxifraga stellaris*. 72 *Sedum rosea*. 73 *Sibbaldia procumbens*.
74 *Silene acaulis*. 75 *Solidago virgaurea*. 76 *Thalictrum alpinum*. 77 *Tofieldia*
pusilla. 78 *Tricophorum cespitosum*. 79 *Trientalis europaea*. 80 *Trisetum*
spicatum. 81 *Vaccinium myrtillus*. 82 *Vaccinium uliginosum*. 83 *Vaccinium*
vitis-idaea. 84 *Veronica alpina*. 85 *Viola palustris*
Source: After R. J. Whittaker (1989)

DCA axis III (λ_3 = 0.181) and DCA axis IV (λ_4 = 0.146) are not isolated within the plexus; they may be interpreted in terms of species composition but seem not to be important. DCA axis I is patently a vegetation gradient. At its positive end cluster heath species including *Betula nana*, *Vaccinium uliginosum*, and *Empetrum hermaphroditum*, while at its negative end cluster pioneer species including *Saxifraga cespitosa*, *Cerastium* spp., and *Deschampsia alpina*. Segregation of species is not so pronounced on DCA axis II. At the positive end species of wet or snow-bed habitats are found (e.g. *Saxifraga rivularis*, *Viola palustris*, and *Pinguicula vulgaris*); while at the negative end are species associated with drier or more exposed location with shorter snow-lie (e.g. *Dryas octopetala*, *Lycopodium annotinum*, and *Silene acaulis*).

Interrelationships within the factor complexes, and in relation to the vegetation axes, were explored by Whittaker (1989) using simple and partial correlation. The terrain-age factor complex has the highest correlation with the DCA axis I, which is also significantly correlated with frost churning, overall disturbance, and, to a lesser degree, with altitude and the three soil variables. However, when controlling for terrain age by taking DCA axis I as the dependent variable and computing partial correlation coefficients, it was found that soil depth, altitude, and disturbance are not significantly correlated with DCA axis I, but frost churning remains significant. The role of altitude in the plexus probably reflects a simple spatial correspondence with terrain age – an altitudinal range of 50 m is unlikely to affect the ecosystem appreciably. DCA axis II is highly correlated with exposure, snowmelt, and moisture. It is also correlated with frost churning, a member of the terrain-age factor complex; this seems to indicate that frost churning is associated with exposed microtopographic sites as well as with young terrain.

Direct gradient analysis was performed using canonical correspondence analysis (Figure 9.6). The first CCA axis is closely related to variables in the terrain-age factor complex – terrain age (τ = 0.92), disturbance (τ = –0.78), frost churning (τ = –0.73), and altitude (τ = –0.73). As before, the most important vegetation axis reflects a successional trend from young, actively disturbed sites with shallow soil (negative scores), to old, inactive sites with deep soils. The second and third CCA axes may be interpreted as character-izing different aspects of the microtopographic factor complex. On CCA axis II, the most important variables are maximum slope angle (+0.46), exposure (+0.39), frost churning (–0.38), litter depth (+0.36), and snowmelt (+0.31). Moisture is the only important variable on CCA axis III.

In summary, the investigations using indirect and direct gradient analysis confirmed the importance of terrain age and associated factors (frost churning, slope movement, and the overall disturbance regime – frost churning, slope movement, grazing, trampling, and stream flow) in explain-ing vegetation distribution. They showed clearly the value of a multivariate approach that tackles many aspects of the environmental complex within which a geoecosystem functions.

Changing disturbance regimes

A change in disturbance regime is likely to affect landscapes. Global warming, should it occur, would probably favour a rise in the rate of forest disturbance owing to an increase in weather conducive to forest fires (drought, wind, and natural ignition sources), convective winds and thunderstorms, coastal flooding, and hurricanes. To simulate the effects of increased disturbance, a FORENA model was modified so that the disturbance probability of all trees on a given plot could be specified (Overpeck *et al.* 1990). Two sets of simulation were run. In the first set ('step function' experiments), simulated forest was grown from bare ground under present-day climate for 800 years so that the natural variability of the simulated forest could be characterized. At year 800, a single climatic variable was changed in a single step to a new mean value, which would perturb the forest, and the simulation was continued for a further 400 years. In each perturbation experiment, the probability of a catastrophic disturbance was changed from 0.00 to 0.01 at year 800. This is a realistic frequency of about one plot-destroying fire every 115 years when a 20-year regeneration period (during which no further catastrophe takes place) of the trees in a plot is assumed. In each of the step-function simulations, three types of climatic change (perturbation) were modelled: a 1 °C increase in temperature; a 2 °C increase in temperature; and a 15 per cent decrease in precipitation. In the second set of simulation runs ('transient' experiments), forest growth was, as in the step-function experiments, started from bare ground and allowed to continue for 800 years under present climatic conditions. Then, from the years 800 to 900, the mean climate, both temperature and precipitation, was changed year by year in a linear manner to simulate a twofold increase in the level of atmospheric carbon dioxide, until the year 1600 when mean climate was again held constant. As in the step-function experiments, the probability of forest disturbance was changed from 0.00 to 0.01 at year 800. In all simulation runs, a relatively drought-resistant soil was assumed, and the results were averaged from forty random plots into a single time-series for each model run.

The model was calibrated for selected sites in mixed coniferous–hardwood forest, Wisconsin, and southern boreal forest, Quebec. Selected summary results are presented in Figure 9.7. Apparently, an increase in forest disturbance will probably create a climatically induced change of vegetation that is equal to, or greater than, the same climatically induced change of vegetation without forest disturbance. In many cases, this enhanced change resulting from increased disturbance is created by rapid rises in the abundances of species associated with the early stages of forest succession

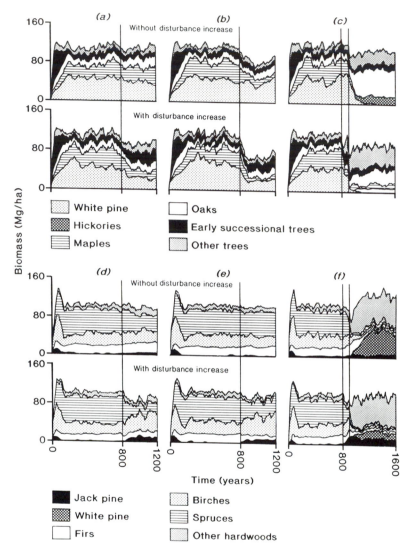

Figure 9.7 Simulated changes in species composition of forests at two sites investigated in eastern North America. (*a*) to (*c*) is a site in Wisconsin, and (*d*) to (*f*) is a site in southern Quebec. At both sites, experiments were run with an increase in disturbance at year 800 (top figure for each site) and without an increase in disturbance at year 800 (bottom figure for each site). Additionally, three climatic change scenarios were simulated: a 1 °C temperature increase at year 800 (left-hand figures, *a* and *d*); a 15 per cent decrease in precipitation (middle figures, *b* and *e*); and a transient change in which mean monthly precipitation and temperature were changed linearly from year 800 to year 900 and thereafter held constant (right-hand figures, *c* and *f*)

Source: From Richard Huggett, *Modelling the Human Impact on Nature: Systems Analysis of Environmental Problems*, 1993, by permission of Oxford University Press, after Overpeck *et al.* (1990)

owing to the increased frequency of forest disturbance. In some cases, as in Figures 9.7a, d, and e, a step-function change of climate by itself does not promote a significant change in forest biomass, but the same change working hand-in-hand with increased forest disturbance does drastically affect forest composition and biomass. Interestingly, the altered regimes of forest disturbance, as well as causing a change in the composition of the forests, also boost the rate at which forests respond to climatic change. For instance, in the transient climate change experiments, where forest disturbance is absent through the entire duration of the simulation period, vegetation change lags behind climatic change by about 50 to 100 years, and simulated vegetation takes at least 200 to 250 years to attain a new equilibrial state. In the simulation runs where forest disturbance occurred from year 800 onwards, the vegetational change stays hard on the heels of climatic change and takes less than 180 years to reach a new equilibrium composition after the climatic perturbation at year 800.

VOLCANIC DISTURBANCE

Disturbance in continental and global landscapes is caused by three different kinds of process: short sharp highly energetic events – bombardment and massive volcanic explosions; protracted bouts of volcanism; and very protracted geological changes associated with plate tectonics. Plate tectonic process, including the drifting of continents and tectono-eustatic changes of sea-level, operate very slowly, cause disturbance over geological time-scales, and are beyond the scope of this book. Volcanism and bombardment are far quicker acting, disturb present-day geoecosystems, and will be considered briefly.

Volcanoes are very common features of the globe (see Francis 1993). There are some 600 active, and several thousand extinct, volcanoes on the continents or exposed above the sea as islands. Submarine volcanoes are legion: at least 50,000 exist on the floor of the Pacific Ocean. This multitude of volcanoes influences organisms and soils indirectly through the relief and parent material factors. Organisms are affected directly, considerable disturbance being caused through volcanic explosions, lava flows, and nuées ardentes, through the leakage of poisonous gases from vents and lakes, and through the release of radioactive substances such as radon.

Local disturbance

Volcanoes cause local devastation and affect immediately surrounding landscapes. The worst devastation results from the volcanic lava, ejecta, and hot gases. This was starkly illustrated by the excavations of Pompeii and Herculaneum which were destroyed and buried by the Vesuvius eruption of 24 August, AD 79. Less dramatic, but equally lethal, are local climatic effects triggered by eruptions. Alexander von Humboldt observed a 'singular meteorological process' to which he gave the name 'volcanic storm': the hot steam rising from a volcanic crater during the eruption spreads into the atmosphere, condenses into a cloud, and surrounds the column of fire and cinders which ascends several thousand feet and generates forked lightning (Humboldt 1849: i, 231). Lightning from the storm triggered by the eruption of Katlagia, Iceland, on 17 October 1755 killed eleven horses and two men! The Laki fissure eruption, which also occurred in Iceland, is known by Icelanders as the Skaftár fires. It involved the eruption of basaltic lava from a chain of 140 craters. The chain ran for some 27 km and cut across an older volcanic ridge called Laki. Starting at 9 a.m. on Whit Sunday 8 June 1783, the eruption lasted eight months, until 7 February 1784. In all, 14 km³ of lava were erupted. About 565 km² of land were buried, including two churches and fourteen farmsteads. The protracted eruption had grave consequences for the Icelandic population. Volcanic gases and aerosols, laden with toxic fluorine, spread across the island. Summer crops and three-quarters of the livestock were killed. The subsequent famine claimed the lives of nearly a quarter of the inhabitants.

Of particular interest in the present context is the relationship between volcanic disturbance and natural geoecosystems. Not much work has been carried out in this field, but a study looking at the survival and recovery of plants following the eruption of Mount St Helens depended on the nature of the volcanic disturbance that the site experienced, and the aspect, time of snowmelt, and ongoing disturbances such as erosion (Adams *et al.* 1987).

Regional disturbance

Volcanoes inject dust and gases into the air. This can change the composition of the atmosphere sufficiently to force a change in regional climates. The indirect influence of such climatic perturbations on landscapes is poorly understood. Climatic changes induced by volcanic eruptions may affect beings living far from the zone of direct damage, but this is not certain. Gilbert White, in his famous *The Natural History of Selbourne* (1789), begins Letter 65 as follows:

> The summer of the year 1783 was an amazing and portentous one, and full of horrible phenomena; for, besides the alarming meteors and tremendous thunder-storms that affrighted and distressed the different counties of this kingdom, the peculiar haze, or smoky fog, that prevailed for many weeks in this island, and in every part of Europe, and even beyond its limits, was a most extraordinary appearance, unlike anything known within the memory of man. By my journal I find that I had noticed this strange occurrence from June 23rd to July 20th inclusive. . . . The sun, at noon, looked as blank as a clouded moon,

and shed a rust-coloured ferruginous light on the ground, and floors of rooms; but was particularly lurid and blood-coloured at rising and setting.... The country people began to look with a superstitious awe at the red, louring aspect of the sun; and indeed there was reason for the most enlightened person to be apprehensive ...

These 'horrible phenomena', with the exception of the 'alarming meteors', were the effects of the Laki eruption. Benjamin Franklin (1784) noted the same effects during a sojourn in Paris.

A classic example of climatic and ecological disruption in the aftermath of a volcanic eruption is the 'Year Without a Summer', also known as the year of the 'Last Great Subsidence Crisis in Europe'. During the year in question – 1816 – the weather in the Northern Hemisphere was downright miserable. It happened to follow the stupendous eruption of Tambora volcano on 10–11 April 1815 (Stothers 1984). This eruption produced about 150 km^3 of ash (the equivalent of 50 km^3 of trachyandesite magma), more than in any other known eruption in the last 10,000 years. The ash cloud rose about 43 km into the stratosphere and spread around the world. Within three months of the eruption, the ash was causing optical effects in the atmosphere over Europe. Several observers in London in late June and early July, and again in September and early October, noticed prolonged and brilliantly coloured sunsets. These sunsets, it is said, inspired many of John Turner's landscape paintings. During the next year, 1816, the summer in western Europe was cool, wet, and gloomy. Crops failed. Famine, disease, and social unrest were rife. In Merionethshire, in Wales, there were reportedly only three or four days without rain between May and October. (That's good going, even for Wales.) It was perhaps no coincidence that during this year, Mary Shelley, forced to stay indoors, wrote a short story that would lead to her novel *Frankenstein*. Lord Byron, spending a wet and wretched summer on the shores of Lake Geneva, penned the sombre poem *Darkness*. His mood is captured in the following extract:

> The bright Sun was extinguish'd, and the stars
>   Did wander darkling in the eternal space
> Rayless and pathless, and the icy earth
>   Swung blind and blackening in the moonless air;
> Morn came and went – and came, and brought no day ...

Weather maps of the time suggest that the Tambora aerosol cloud caused the extraordinary climatic conditions. In the North Atlantic region, the cloud led to a prolonged and marked drop in surface pressure across the mid-latitudes. This caused the travelling westerly depressions to adopt a more southerly route, so producing a miserably cold and wet summer over Europe. North America did not escape the climatic perturbations caused by Tambora. The Sun was reported to be dim in the north-eastern United States during the year 1816, owing to a persistent 'dry fog' that was dispersed by neither wind nor rain. In New York, according to one observer, the Sun was so reddened and dimmed by the fog that sunspots could be seen by the naked eye. The Hudson's Bay Company trading posts recorded the coldest summers in modern records in the years 1816 and 1817. Frost-ring records from trees in central western Quebec support these observations. The atmospheric circulation pattern in North America at this time, as reconstructed by the distribution and severity of sea-ice in the Hudson Strait in 1816, allowed Arctic air to make southwards incursions into eastern North America and eastern Europe. This would account for the unseasonably cold weather during the spring and summer of 1816 in eastern Canada and the north-eastern United States. Repeated frosts and snow in late spring and summer caused crop failures, poor harvests, and famine in New England.

The global effects of Tambora are manifested in several climatic indicators from around the world. Tree rings register frost damage from the western United States to South Africa. The grape harvest was very late in France, indeed the latest it had ever been in four centuries. The forests on the island of Trindade, Brazil, appear to have been destroyed. The most significant change seems to have been the shortening of the growing season in the Northern Hemisphere. There were, as has been seen, also episodic outbreaks of severe weather. These two climatic changes account for the calamitous agricultural failures of 1816.

The effects of older volcanic eruptions are more difficult to gauge, but written and archaeological sources point to eruptions associated with falls of temperature and crop failures in Classical times (Stothers and Rampino 1983). Egyptian and Biblical writings report darkness and rains of ash at the time of the Exodus, and a coincidence between these unusual conditions and the explosive eruption of Santorini (Thera) in the Aegean Sea in the second millennium BC has been noted (e.g. D. J. Stanley and Sheng 1986). The explosion of Santorini may have contributed also to the downfall of Minoan Crete (Marianatos 1939), and have had environmental repercussions as far away as China (Pang and Chou 1985). Less dramatic changes may occur as well. Frost damage in the annual rings of *Pinus longaeva* and *Pinus aristata* exhibits a good relationship with the timing of known volcanic events (LaMarche and Hirschboeck 1984). Studies of tree rings in Irish bog oaks seem to reveal volcanic events at 4375 BC, 3195 BC, 1628 BC, 107 BC, and AD 540, which dates may correspond to peaks of activity in Greenland ice cores (Baillie and Munro 1988).

Global disturbance

The largest eruptions, termed 'supereruptions', may have global consequences, producing 'volcanic winters' similar to the recently proposed 'nuclear winters' (Rampino *et al.* 1985, 1988) and 'cosmic winters' (Clube and Napier 1990). To be sure, modern volcanic eruptions are dwarfed by some in the recent past. Toba, in Sumatra, exploded 75,000 years ago, and is the world's largest known eruption (Rose and Chesner 1990). About 2,800 km^3 of dense rock equivalent were erupted as a tuff sheet covering 20,000 to 30,000 km^2; 800 km^3 of ash were deposited as a blanket over the Indian Ocean and southern Asia; and an estimated 1,000 to 5,000 megatonnes of sulphuric acid aerosols were lofted into the atmosphere. This compares with 100 megatonnes for Tambora. Whereas Tambora caused a dimming of the Sun and the 'Year Without a Summer', Toba would probably have precipitated a volcanic winter in which photosynthesis stopped and wildfires occurred causing severe disturbance to the biosphere. Such supereruptions might have played a role in the initiation and timing of glacial–interglacial cycles, and set in train a new megascale disturbance regime.

In the remoter past, changes in climate produced by lengthy periods of intense volcanism may have disturbed the biosphere severely enough to cause mass extinctions. This was possibly the case for the Cretaceous event (Officer *et al.* 1987; Officer 1993). The close of the Cretaceous period saw a paroxysm of intense volcanicity: flood basalts poured over large parts of India, and possibly over the North American Tertiary Igneous Province which appears to have been active at the same time as the Deccan Province (McLean 1981; Courtillot and Cisowski 1987). This volcanism would have created large volumes of dust and gases that, through climatic changes, could have disturbed the biosphere. Particularly disastrous would have been the large injections of sulphates into the atmosphere: large amounts of acid rain would have fallen, the alkalinity of the surface ocean would have dropped, the atmosphere would have cooled, and the ozone layer would have been depleted (Stothers *et al.* 1986). There seems little doubt that the Deccan Province developed rapidly during an episode of immense volcanism that coincided with the Cretaceous–Tertiary boundary and important floral and faunal changes (Duncan and Pyle 1988; Courtillot *et al.* 1988). In detail, the province grew by a series of individual eruptions separated by repose periods. Each eruption would undoubtedly have caused dramatic environmental deterioration and disturbed the biosphere, while repose periods would have provided the biosphere with a chance to recover. The sheer magnitude of the volcanism in this province is not in itself sufficient to explain irreversible changes in the fauna and flora: the eruptive rate and length of repose periods are equally important factors (Cox 1988). It is difficult to assess the number of eruptions in a volcanic province, but, from field studies in the Deccan, it is estimated that there were something between 100 and 500 eruptive events (Cox 1988). The entire period of volcanism lasted about 500,000 years, so the average repose period is between 1,000 and 5,000 years. As there are no modern analogues of flood-basalt volcanism, it is tricky to gauge the significance of these numbers to biospheric recovery. One can only conclude that volcanic activity might precipitate mass extinctions through a series of large eruptions over thousands of years (e.g. Courtillot 1990); the chances of a mighty one-off explosion having the same effect are slim. However, the evidence for massive bouts of volcanism is mounting as the full extent of large igneous provinces, remnants of geologically brief phases of magmatic activity, becomes apparent (Coffin and Eldholm 1993). Is it a coincidence that the Siberian traps flood basalts, one of the most voluminous flood basalt provinces on the Earth, erupted over an extremely short time span at the close of the Permian period (Renne and Basu 1991), at the time of a spectacular mass extinction?

COSMIC DISTURBANCE

Space exploration and space science have led to a quite revolutionary conclusion: throughout its entire history the Earth has been hit by 'stray' cosmic bodies. The Earth's surface is dotted with the remains of craters formed by the impact of asteroids, comets, and meteorites. Over 120 craters have so far been discovered which show signs of having been produced by the impact of an extraterrestrial body. Impacting bodies not only excavate craters, they also set in train a sequence of events that causes change in the ecosphere, lithosphere, and barysphere. The bombardment hypothesis is having an enormous impact on the Earth and life sciences. In the context of the 'brash' equation, bombardment is a driving variable of the disruptive kind that, given a large enough impactor, will profoundly disturb geoecosystems at all scales, and will destroy a sizeable part of the biosphere. Smaller impactors will cause less direct damage to the biosphere, but may bring about radical changes in mesoscale and macroscale landscapes.

Nobody has witnessed a large strike in modern historical times. The biggest and most well-documented encounter between an Earth-crossing body and the Earth was the Tunguska event of the morning of 30 June 1908. A stony meteorite, travelling from south-east to north-west over the Podkamennaya–Tunguska River region of Siberia, produced a great fireball about 60 km north-west of the remote trading post of Vanovara. The fireball, which exploded about 10 km above the ground, could be seen 1,000 km away and the atmospheric shock heard at even greater distances. Trees within a 40 km radius of the blast were knocked flat, and there is some evidence that dry timber was ignited. The energy released by the explosion is estimated to have been 10 and 20 megatonnes of TNT equivalent energy (Chyba *et al*. 1993), roughly the same energy as a very large hydrogen bomb. This is the largest event for which there is direct evidence. But space probes have shown that impact craters are common on other planets and satellites throughout the Solar System. The magnitude and frequency of impact events can be calculated using the size-distribution of these craters on other planets and satellites.

The frequency of collision with planetesimals is inversely proportional to the size of the colliding body. Meteoritic dust continually enters the atmosphere. Larger meteorites strike on an annual, decadal, centennial, and millennial basis. Asteroids with a diameter of about a kilometre strike about thrice every million years. Mountain-sized asteroids or comets strike about once every 50 million years. At present, several hundred asteroids and cometary nuclei with diameters greater than 100 m have been observed in Earth-crossing orbits. Extrapolation of this known population of near-Earth objects indicates that about 2,000 objects with diameters greater than 1 km possess Earth-crossing orbits. Methods of deflecting oncoming bolides are currently being assessed, such is the grave concern over the effects that a collision with even a small object might produce (Ahrens and Harris 1992).

Single impacts

Since at least the seventeenth century, it has been recognized that a close encounter, or indeed a collision, between the Earth and a comet would have disastrous consequences (see Bailey *et al.* 1990). Not until the 1940s and 1950s was the idea taken up eagerly. The view that a random impact might have global consequences was vented by the astronomer Harvey Harlow Nininger in 1942. Nininger believed that a collision between the Earth and a large meteorite would have several consequences: great changes in shorelines, the elevation and depression of extensive areas, the submergence of some low-lying areas of land, the creation of islands, withdrawal and extension of seas, widespread and protracted volcanism, and the sudden extinction of biota over large areas. Subsequently, connections between bombardment and extinction events were suggested (see de Laubenfels 1956; Öpik 1958; McLaren 1970; Urey 1973). None of these suggestions was taken very seriously, probably because, interesting though they were as speculations, they could not be tested.

Recently, a much fuller appreciation of the process of bombardment and its potential disrupting effects on the biosphere has been gained. This has been made possible by improved information on mass-extinction patterns and processes, and the discovery of what seems to be post-impact fallout in the stratigraphical column, especially at the Cretaceous–Tertiary boundary (Alvarez *et al.* 1980; Alvarez 1986). A spur to this frenzied activity was given by Walter Alvarez and his colleagues. In 1979, they discovered a marker horizon at the boundary between the Cretaceous and Tertiary periods. They took this horizon as concrete evidence that the extinction event at the close of the Cretaceous period was indeed geologically instantaneous and caused by an asteroid colliding with the Earth (Alvarez *et al.* 1980). Their most publicized find was made in the Bottaccione Gorge, near the medieval town of Gubbio, in the Italian Apennines. The gorge cuts through a section of Upper Cretaceous and Lower Tertiary limestones. At the Cretaceous–Tertiary boundary, a distinctive layer of clay is sandwiched between the pelagic limestones. The clay layer contains anomalously high concentrations of iridium, concentrations much greater than could be explained by the normal rain of micrometeorites. Similar anomalously high concentrations of iridium were soon found at Cretaceous–Tertiary boundary sites near Højerup Church, at Stevns Klint, Denmark and near Woodside Creek, New Zealand, and at other sites around the world. The iridium anomalies were inconsistent with a reduction in sedimentation rate of the other components in the clay, and they were unlikely to have come from a supernova. It occurred to Luis W. Alvarez in 1979 that the iridium might have been spread world-wide by a dust cloud thrown up by the impact of an Apollo asteroid measuring about 7 to 10 km in diameter. The cloud of dust would have broadcast iridium from the asteroid around the globe, and by causing several months of darkness, would have led to a collapse of ecosystems and mass extinctions. Subsequent investigations showed that the iridium anomaly is a world-wide phenomenon (Alvarez *et al.* 1982), a fact which, at first sight, seemed to vindicate the notion of an extraterrestrial event at the Cretaceous–Tertiary boundary. However, it has been argued that volcanism might account for the features attributed to an impact event. The matter is still being debated.

The impact of large bolides would cause both instant and delayed disturbance to landscapes. Within seconds of piercing the atmosphere, a mountain-sized bolide would create a superwind capable of flattening forests within 500 to 700 km (Emiliani *et al.* 1981). The impact itself would create a blast wave producing overpressures capable at their peak of destroying forests and killing animals (Napier and Clube 1979). Particularly vulnerable would be large land vertebrates with a small strength–weight ratio. This fact has been used to explain the selective extinction of large dinosaurs during the Cretaceous event (Russell 1979). A wave of intense heat would also radiate from the site of impact, killing all exposed life-forms within the lethal radius. For impacts of bodies 10 km in diameter, the lethal radius could encompass landscapes of continental size. The intense heat associated with a large-body impact may set off wildfires, so releasing soot into the atmosphere (Wolbach *et al.* 1985, 1988). The mechanisms by which wildfires might be ignited are somewhat debatable, but thermal radiation generated by ballistic re-entry of ejecta into the atmosphere could have caused a 150-fold increase in the global radiation flux, over the normal input from solar radiation, for one to several hours (Melosh *et al.* 1990). Thermal radiation inputs of this magnitude may well have been responsible for sparking off wildfires as well as directly damaging exposed animals and plants. The extra mass added to the atmosphere by an impact would cause within hours a rise of mean global surface air temperature of about 1 °C, the maximum rise occurring near the impact site. The great temperatures would encourage large quantities of nitrogen oxides to form. A very large impact could produce up to 3×10^{18} g of nitric oxide which, in less than a year, would spread through the atmosphere to give a world-wide, atmospheric nitrogen dioxide concentration of 100 ppm by volume – a level a thousand times higher than during the worst air pollution episodes in modern cities (Prinn and Fegley 1987). Large particles would fall out rapidly, but a fine cloud of dust would spread globally. The dust cloud would stay in suspension for months or years, blocking out sunlight (Toon *et al.* 1982), and causing dramatic, though patchy, cooling at the surface (Covey *et al.* 1990). The darkness would lead to a reduction or collapse of photosynthesis (Alvarez *et al.* 1980; Emiliani *et al.* 1981) and a breakdown of food chains (Russell 1979).

The impact hypothesis of the Cretaceous extinction event has been given a fillip by the discovery of the 'smoking gun' – a crater at Chicxulub on the Yucatán Peninsula, Mexico, that is of just the right age (e.g. Hildebrand *et al.* 1991). A dramatic change in climate, whatever its precise nature, might explain the relatively sudden disappearance of large parts of the fauna and flora at the close of the Cretaceous period. Radical climatic changes caused by the same mechanism might have occurred at the Precambrian–Cambrian (McLaren 1988b), Devonian–Carboniferous (Wang *et al.* 1991), and Permo–Triassic boundaries (Hodych and Dunning 1992; McLaren 1988b), and possibly the Ordovician–Silurian boundary as well (Wang *et al.* 1992). The evidence is not equivocal, but the indication is that bombardment has given the biosphere a sharp nudge on several occasions (e.g. D. M. Raup 1990; Sharpton and Grieve 1990).

The effects of impact episodes

The impact of a large object is a relatively rare, probably random, event. Occasional collisions of the Earth with kilometre-sized asteroids and cometary fragments occur once every 100,000 to 1,000,000 years or so. As mentioned earlier, these infrequent encounters will jolt the biosphere considerably. Collisions with debris less than about a kilometre in diameter, though far less energetic, are much more frequent, much better co-ordinated, and still have a significant impact on the biosphere. A current astronomical theory proposes that small impacting objects arrive, not singly and randomly, but in epochs of high activity and as a coherent set. The suggestion is that large or giant cometary bodies, with diameters of 100 km or more, arrive randomly, roughly once per 100,000 years, in short-period orbits with perihelia in the inner Solar System. There seem to be two significant sources of giant comets in the inner Solar System producing Jupiter-family comets and Halley-type comets (Bailey *et al.* 1994). Jupiter-family comets are normally returned to orbits with perihelia close to Saturn within 10,000 years. They possibly return some five to ten times at mean intervals of about 100,000 years. On the other hand, Halley-type comets probably persist in the inner Solar System for about a million years. Such large comets are likely to disintegrate owing to thermal stress while at and near perihelion, collisions in the asteroid belt, and gravitational stresses within the Roche limit of Jupiter.

Giant comets supply copious quantities of dust and interplanetary debris throughout their short-period phases. Their disintegration produces streams of high-velocity material in Earth-crossing orbits. The streams are clusters of kilometre-sized objects in short-period orbits and appearing as Apollo-type asteroids, smaller objects with diameters in the range 50 to 300 m, and micron-sized dust. They will have repetitive intersections with the Earth when precession brings the node around to 1 AU, and when the cluster passes its node when the Earth is nearby. This leads to impacts at certain times of year, every year, during active periods. The active periods last about one or two centuries, and occur once every few millennia. They could produce lengthy, major episodes of deterioration in the Earth's near-space environment. One such cluster of material was formed by the fragmentation of a giant comet that arrived in the inner Solar System some 20,000 year ago and is still orbiting there now. This giant comet was probably a progenitor of comet Encke and the Taurid meteor showers. Over the last 20,000 years it has produced episodes of atmospheric detonations and stratospheric dusting with significant consequences for the terrestrial environment and for humankind. It may have been the principal cause of climatic variations during the Holocene epoch and been the culprit responsible for bringing down many great civilizations by triggering cosmic winters (Clube and Napier 1990; Asher and Clube 1993) (Figure 9.8).

It is possible that exceptionally long-lived bouts of terrestrial bombardment in the geological record are associated with the evolution and decay of Halley-type giant comets (Bailey et al. 1994). This would be more likely if the break-up of the comet whilst in a short-period sun-grazing orbit coincided with the debris stream intersecting the Earth's orbit. One can envisage a dense stream of material containing meteoroids and cometary debris ranging in size from sun-grazers tens or hundreds of metres across (comparable in size to the Tunguska projectile), up to kilometre-sized cometary nuclei. This stream might produce long-lived environmental stress on the Earth through both repeated impact events, similar in principle to 'comet showers' (Hills 1981), and stratospheric dusting. Some of the mass extinctions in the geological record might have resulted from the fragmentation of a giant comet in a short-period orbit that produced a massive debris stream intersecting the orbit of the Earth (Bailey et al. 1994). This pattern of bombardment does seem to accord better with the fossil record than a one-off impact scenario: investigations of many boundary sites have indicated that mass extinctions occurred, not instantaneously, but in a series of discrete steps spread over a few million years. However, although there is normally a protracted change in biota prior to supposed extinction events, there is 'in every case a single horizon, recognizable globally, at which the main biomass disappeared, bioturbation ceased, and (for a variable time) an almost total

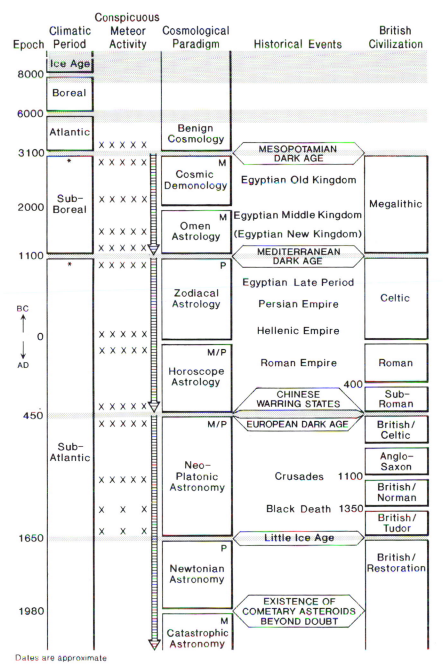

Figure 9.8 Historical meteor activity, climate, and the rise and fall of civilizations
Source: From Clube and Napier (1990)

loss of life occurred' (McLaren and Goodfellow 1990: 161; cf. McLaren 1988a). And, whatever caused the terminal Cretaceous extinction event, it appears to have acted uniformly over the globe: marine bivalves show no regional or latitudinal gradients of extinction (D. M. Raup and Jablonski 1993).

SUMMARY

Geoecosystems at all scales are subject to disturbance. Disturbing agencies may come from within the ecosphere, from inside the Earth, and from space, to create ecological disturbance, geological disturbance, and cosmic disturbance, respectively. Ecological disturbance is effected by biotic and abiotic (physical) agencies. Abiotic agencies include strong winds, which lead to tree-throw, floods, and fire. Biotic agencies include pests, pathogens, and grazing. Recent work on disturbance focuses on patch dynamics and on disturbance regimes, where several disturbing agencies act in concert. Many communities appear to adapt to disturbance regimes. Multivariate methods can be used to assess the relative significance of disturbance and other environmental factors to community structure and composition. Given that the environment, and especially climate, appears to be ever changing, it is reasonable to suppose that disturbance regimes will also change. The possible effect of changing disturbance regimes associated with a warming world have been investigated using a forest growth model. Geological disturbances act over all time-scales. At times during human history, volcanic disturbance has been considerable. All volcanoes create local disturbance. Regional disturbance is not uncommon, but depends on many factors, including the nature of the eruption and the composition of the ejecta. The most recent major disturbance followed the eruption of the Tambora volcano in April 1815. Massive global disturbances follow in the aftermath of supereruptions, as when Toba, Sumatra, exploded some 75,000 years ago. The massive outpouring of flood basalts over prolonged periods of time may also have had apocalyptic consequences for the ecosphere. Over the last couple of decades, it has become accepted that the Solar System and Galaxy are violent places. Cosmic disturbance ranges from stratospheric dusting with micron-sized meteoritic particles, to a collision with a mountain-sized comet or asteroid. Single impacts with massive bolides have almost certainly occurred, and may have caused mass extinctions. More modest impacts, mostly involving heavy stratospheric dusting and atmospheric detonations, would still have dire consequences for life. It is possible that climatic change over the last 20,000 years has been driven by periodic collision with cometary material, the debris from a giant progenitor of the comet Encke and the Taurid meteor showers. These collisions might have created cosmic winters and brought about the fall of civilizations.

FURTHER READING

Disturbance of ecosystems is a current focus of interest and is widely discussed in the ecological and geoecological literature. A feel for the basic ideas may be gained by dipping into the following books: *The Ecology of Natural Disturbance and Patch Dynamics* (Pickett and White 1985) and *Landscape Heterogeneity and Disturbance* (M. G. Turner 1987). Some recent papers look at disturbance over different spatial scales (e.g. M. G. Turner *et al.* 1993). Volcanic disturbance is widely researched but information is rather scattered. Peter Francis's book *Volcanoes: A Planetary Perspective* (1993) contains a valuable chapter, called 'The golden glow of volcanic winter', dealing with the effects of volcanoes. The case of Mt St Helens is documented in the *United States Geological Survey Professional Paper* entitled 'The 1980 eruptions of Mt. St. Helens, Washington' (Lipman and Mullineaux 1981). Cosmic disturbance has been headline news in the academic and popular press since the famous iridium layer was first reported by Alvarez and his colleagues in 1980. The volume called *Global Catastrophes in Earth History; An Interdisciplinary Conference on Impacts, Volcanism, and Mass Mortality* (Sharpton and Ward 1990) provides a good summary of the debate. It is worth reading up on the Chicxulub crater (e.g. Hildebrand *et al.* 1991). *The Cosmic Winter* by Victor Clube and Bill Napier (1990) makes interesting reading, though it is perhaps better to read *The Origin of Comets* (Bailey *et al.* 1990), especially the chapter dealing with catastrophic impacts, first. Derek V. Ager's (1993) comments of catastrophes made in a chapter entitled 'Jupiter or Pluto?' in his *The New Catastrophism* (1993) are worth considering. This field of terrestrial catastrophism is still developing fast. Most new findings are reported in *Nature, Science,* and *Geology.*

EPILOGUE

This book has essayed three tasks. First, it has presented and tried to justify a broader view of the landscape sphere than is taken currently, proposing a hierarchy of geoecosystems. Second, it has attempted to formulate within a dynamic systems framework a general model of geoecosystems, expressed as the 'brash' equation, that captures ecological and evolutionary aspects of geoecosystems. This approach owes much to Jenny's inspirational 'clorpt' equation. Third, it has endeavoured to show how, as a conceptual tool, the model is able to probe the many-faceted environmental influences on landscapes. To finish, conclusions will be drawn about the merits and demerits of the proposed approach.

THE LANDSCAPE AS A HIERARCHY OF INTERDEPENDENT SYSTEMS

The suggestion that the landscape might profitably be viewed as a hierarchy of nested systems is unlikely to cause a rumpus since it is fashionable to think of Nature in terms of interdependence within hierarchical structures. Indeed, many people see Nature as a vast edifice that is continually organizing and reorganizing itself on hierarchical lines (e.g. Laszlo 1993).

The rise of global ecology and the Gaia hypothesis has led to the notion that all terrestrial phenomena are interrelated. Various schools of thought on the nature of this interdependence exist, representing all shades between stark mechanism and pure vitalism (see Kirchner 1991). This global picture of the ecosphere is important to the 'brash' equation since it is a supersystem within which all geoecosystems operate. Ecologists are now addressing environmental problems that require understanding at all scales, from microscale to megascale. The gargantuan problem is to model processes across these scales, something not discussed in this book, primarily because most work on this topic is theoretical. A feel for the enormity of the task can be gleaned from considering the scales involved in modelling climatic change where ecological and atmospheric submodels include processes that occur at the scale of a leaf and influence megascale atmospheric processes and patterns. To predict across scales, it is necessary to identify processes of interest, define suitable parameters affecting the processes at different scales, develop rules to translate information across scales, and devise methods to test these predictions at the relevant spatial and temporal scales (M. G. Turner *et al*. 1989). It is an area that must surely warrant thorough exploration in the future, if for no other reason that what goes on in mesoscale and macroscale landscapes, which the human species are transforming rapidly, affects the ecosphere as a whole.

In proposing to widen the scope of geoecosystems from the confines of the mesoscale landscapes so dear to the landscape ecologists, I have had the temerity to tidy some rather opened-ended definitions of terms and to invent some new ones. Apologies are extended to all those who prefer to see the biosphere as something more than the sphere of life.

AN ECOLOGICAL APPROACH

The 'brash' equation is ecological in that it describes the interdependencies within geoecosystems. These interdependencies are set down in the 'brash' equation. Many of them have been discussed in the main part of the book. Not all possible interdependencies within geoecosystems were considered, mainly through lack of space. Nothing was said about interactions among organisms, nor much noted about biotic influences on the physical landscape and atmosphere. These are important areas of study that have generated a vast literature. Interactions between organisms have been investigated using a dynamical systems formulation ever since Alfred James Lotka (1925) and Vito Volterra (1926, 1931) set down their classic equations describing predator–prey relationships. The rise of non-linear dynamical systems theory has led to a flowering of Lotka's and Volterra's seminal ideas on population interaction (e.g. May 1976; Tilman and Wedin 1991; J. C. Allen *et al*. 1993; Hanski *et al*. 1993). A new field has emerged – population interactions in heterogeneous environments; in other words, spatial and scale effects are now often considered in dynamical systems models (e.g. Levin 1974, 1976, 1988, 1989, 1991; Levin *et al*. 1989).

Biotic effects on the physical landscape are now seen as highly significant by physical geographers (Viles 1988). The influence of biota on geoecological processes has become a focus of attention. Vegetation is a key factor in many slope development and soil erosion models (e.g. Kirkby 1989), while the importance of animal activities in denudation (e.g. D. S. G. Thomas 1988; Butler 1992) and soil mixing (D. L. Johnson 1990) has recently been emphasized. The effect of vegetation on mesoscale and macroscale climate has been found to be surprisingly great. Simulation using a global climate model shows that boreal forest keeps winter and summer air temperatures warmer than would be the case if the forest were replaced by bare ground or tundra vegetation (Bonan *et al*. 1992). This kind of study overturns the traditional view of climate and vegetation distribution in which the extent and nature of biomes are determined by climate. The basis of this view was the excellent match between, for instance, climatic indices and the boundaries of the boreal forests and its subdivisions (e.g. Bryson 1966). But it now seems that the climate–vegetation correlations are the result of coupled dynamical interactions in which the geographical distribution of boreal forest affects climate and vice versa. Most of the 'traditional' work expressed the role of climate in too simplistic a way. Climatic controls at macroscales and megascales seem undeniable, but are more subtle and complex than previously thought. Also, it was indicated in the book that disturbing agents, many of them climatic, are strong influences on communities. But, rather than being seen as arresting factors that stem the successional sequence of a community to its climax state, disturbing agents should be regarded as an integral part of the system – a community adjusted to the passing of hurricanes is just as much in a steady state as a climax community adjusted exclusively to temperature and precipitation.

Ideally, the 'brash' equation would be explored through dynamical systems analysis. This application has not been mentioned much in the book, largely because there are few examples of it concerned with geoecosystems. The model built to analyse factors controlling soil organic matter levels in grasslands (Parton et al. 1987) illustrates the pragmatic value of a dynamical systems approach, while its theoretical value is evidenced by Phillips's demonstration that soil variability may arise from chaotic dynamics in soil systems (see also Phillips 1993a, 1993b). Other examples of systems models of geoecosystems are discussed elsewhere (e.g. Sklar and Costanza 1991; Huggett 1993). Dynamical systems equations are extensively used to model interactions within the biosphere (as mentioned above), and have been fruitfully applied to geomorphological systems (e.g. Culling 1988; Malanson et al. 1990; Malanson et al. 1992; Phillips 1993c). Our understanding of geoecosystems would surely benefit from dynamical systems analysis because it accommodates non-linearities, discontinuities, and deals explicitly with change, all of which are difficult to make sense of using other methods. An alternative, and perhaps less satisfactory way, of exploring multivariate interactions suggested by the 'brash' formula is to use statistical methods. Several studies mentioned in the book did adopt this approach to good effect. However, it is worth raising two questions about the application of multivariate statistical methods to geoecosystems. First, what is the best strategy to employ? And second, how should the environmental complex be characterized?

Robert J. Whittaker's investigation of vegetation–environment relationships on the Sorbreen glacier foreland helps to answer these points. The indirect and direct gradient analysis confirmed the importance of terrain age and associated factors in explaining vegetation distribution, but slightly different configurations of relations were indicated between the vegetation and microtopographic factors. The two techniques seemed to produce closely comparable results where relations were clear and strong (as in the terrain-age factor complex), but generated differences where relations were more complex and weaker. In the latter case, CCA teased out distinct effects of the interwoven variables, whereas the NMDS plexus approach appeared to furnish a sharper picture of the overall structure and complex of interactions. On the other hand, a CCA analysis omitting sites beyond the glacier foreland boundary produced a result in which the three key variables of the microtopographic factor complex were the three environmental variables most strongly related to CCA axis II (R. J. Whittaker 1989). So, some of the differences between the two techniques may result from the quality of data, especially as CCA demands rigid statistical assumptions. Considering these findings, it would seem that using several methods is the best strategy to adopt in applying multivariate methods to geoecosystems. Many researchers already do this, and it is a practice greatly to be encouraged, for it provides the best available guide to patterns in Nature when mathematical models are not available or appropriate.

The second question that was posed concerned how best to characterize the environmental complex. Whittaker used twelve environmental variables, but many of the studies discussed in this book used far fewer than that. The number of variables used to define the environment seems to vary between disciplines. Zoologists are very good at extracting multivariate axes from a large suite of measurements on individual organisms, but seem content to describe the environment by two variables – often mean annual temperature and rainfall. Physical geographers usually describe the environment more fully, and may have something to teach biological scientists. In fairness to zoologists and others who characterize the environment by a couple of variables, climatic variables in macroscale landscapes can often be reduced to two master driving variables – actual evapotranspiration and temperature. However, there seems little point in using multivariate techniques in essentially bivariate situations. It is far better, and will possibly be more rewarding, to define the environmental complex as fully as possible. John A. Matthews (1992: 1) contends that, to progress, ecologists must simplify. While appreciating what is meant by this statement, benefits may accrue from facing complexity head on and trying to establish a more comprehensive description of the environmental complex when studying geoecosystems. In the past, a difficulty with describing the environment using a multitude of descriptors has lain in the acquisition of data. Fortunately, the recent availability of remotely sensed images and the construction of geographical information systems has helped enormously in surmounting this problem. None the less, there is still a vital need for researchers to go into the field to acquire information and ideas.

AN EVOLUTIONARY APPROACH

It is often claimed these days that scientific ideas mirror the cultural and social milieu in which their authors are nurtured. Evolutionary views of life, the universe, and everything are fashionable. The suggestion that geoecosystems may be profitably viewed in an ecological and evolutionary framework is part of a wider move towards emphasizing ever-continuing change in evolving hierarchies (e.g. Laszlo 1993). The 'brash' equation is evolutionary in outlook, and stands in contradistinction to the developmental paradigm of soils and vegetation that emerged towards the close of the last century. From an evolutionary perspective, geoecosystem stability is ephemeral, geoecosystem change is the keynote. Several examples of this evolutionary view in soils and biotas have been considered in the book.

An evolutionary view of soils, as first developed by Donald Lee Johnson and his colleagues, has deep implications. It means that to speak of 'degree of soil development' is misleading. A far more satisfactory term is 'soil state'. Of course, anisotropy is a measure of soil state, but it does not necessarily correlate with age – the steady-state tropism of a soil will vary with landscape setting. A soil on a steep slope may be just as much adjusted to environmental factors as a so-called mature soil on a flat, well-drained site. And within soil groups, the enormous spatial variability might reflect the non-linear and chaotic behaviour of the soil system, rather than random noise. (Noise is a useful term for disguising lack of understanding.) The environment may be thought of as a set of interacting fields that themselves interact with geoecosystems, including the soil. It is nigh on impossible to isolate one environmental factor and extract from the complex web of interactions in geoecosystems a simple relationship between environmental cause and pedogenetic effect, though this can be done to a limited extent under certain circumstances, particularly where the soil property concerned responds rapidly to environmental change (e.g. nitrogen and humus contents). Multivariate systems must be studied using multivariate methods. The majority of pedologists have seemed reluctant to use multivariate methods, except for classifying soils. More generally, it has been suggested in the book that pedogenesis should be viewed in the context of the evolving geoecosphere. This idea was posited in the notion of soil–landscape systems, but has been extended in this book to include animals and plants.

An evolutionary view of animal and plant communities is advocated. As with soils, this means a rejection of the developmental climax concept in favour of a concept that stresses the dynamism of species and communities (cf. Worster 1990). The notion is that all units in the societary hierarchy constantly respond to ever-changing circumstances in their biotic and abiotic environments. If steady states exist in species and communities, then they are temporary balances that will shift as environmental factors change. At the species level, much evidence hints at phenotype responsiveness to environmental influences, at all scales, from small islands to the entire terrestrial globe. At the level of communities, responsiveness is reflected in the inconstancy of species composition – there is nothing permanent about biomes. As an example it is instructive to consider a late Pleistocene biome which existed in north-central United States from about 18,000 to 12,000 years ago. The vegetation of the area was rich in spruce and sedges, and formed a spruce parkland or 'boreal grassland' (Rhodes 1984), perhaps similar to the vegetation found in the southern Ungava Peninsula, northern Quebec, today. The fauna of this biome, as well as the flora, was disharmonious: species which now inhabit grassland or deciduous woodland, including prairie voles, sagebrush voles, and the eastern chipmunk, lived cheek-by-jowl with species that now occur in boreal forests and Arctic tundra, including Arctic shrews, lemmings, voles, and ground squirrels (Lundelius *et al.* 1983). The boreal grassland biome has no modern analogue. This does not mean that the animals and plant species which comprised it were not in harmony with the prevailing Pleistocene environments; it simply means that they were maintained by climates which have no modern counterparts (Graham and Mead 1987: 371).

The uniqueness of communities is also highlighted by an unusual kind of habitat island that occurs in the Bitterroot Mountains, which run north and south for some 80 km on the Montana–Idaho border. This mountain range is dissected by deep valleys cutting in from the east. Since the valleys are very close together and formed in the same rocks, environmental conditions in each are almost identical. For this reason, the communities they contain would be expected to be very similar. The fact is that they are different. Furthermore, the differences cannot be accounted for by any environmental factor, not even a simple latitudinal gradient. So why do they differ? The answer appears to lie in the fact that the valleys are isolated from one another by the intervening mountain ridges. Propagules from one valley cannot cross a ridge to another. This means that species cannot colonize new valleys. The constraints are so severe that no unique sort of equilibrium between the vegetation and the environment evolves. Rather, each valley has its own vegetation arising from the happenstance of what was available at crucial times of establishment and a unique disturbance history (McCune and Allen 1985a, 1985b).

Dynamical geoecosystems almost certainly are replete with non-linearities. Non-linear dynamics contains evolutionary and holistic elements. It has been found that large interactive systems, comprising millions of elements, naturally evolve towards a critical state in which a minor event leads by way of a chain reaction to a catastrophe affecting any number of elements in the system (Bak *et al.* 1988; Bak and Chen 1991). This critical state appears to be poised at the edge of chaos (Kauffman 1992; Lewin 1993). Notions of self-organized criticality and systems at the edge of chaos may explain the dynamics of many phenomena, including ecosystems. A key point is that the processes leading to the minor events are the same ones leading to catastrophes. And, because the global features of the system, such as the relative number of small and large events, do not depend on microscopic mechanisms, the concepts are truly holistic. Now, in a fully chaotic system, a small initial uncertainty grows exponentially with time. However, by running simulations of systems at their critical state, the uncertainty increases according to a power law and not an exponential law. In other words, the system evolves on the border of chaos, and its behaviour, known as weak chaos, is the consequence of self-organized criticality. A fundamental difference between fully chaotic systems and a weak chaotic system is that long-term predictions are possible in the latter case. An example is furnished by avalanches in a sand pile (Bak and Chen 1991). When a grain of sand is added to a sand pile in a critical state it can start a chain reaction leading to an avalanche of any size, including a 'catastrophic' event, but, most of the time, will fall so that no avalanche occurs. Experiments and simulations show that even the largest avalanche involves only a small fraction of the sand grains in the pile and cannot therefore cause the slope of the pile to deviate greatly from its critical angle.

Self-organized criticality emerges from the 'game of life', a computer model that simulates the evolution of a colony and mimics the generation of complexity in Nature (Bak et al. 1989). Essentially, the game starts with randomly distributed organisms in a homogeneous landscape and allows them to interact, produce offspring, and die according to certain rules. The outcome is normally a simple periodic state containing stable colonies. However, if the game is perturbed by adding extra live cells, the system behaves transiently for long periods. Per Bak and his colleagues kept disturbing the game once it had reached a stable pattern, and found that the total number of births and deaths in the 'avalanche' following each additional perturbation followed a power law. This suggested that the system had organized itself into a critical state. Models of this kind may have ramifications for geoecosystems. Forest fires, for instance, seem to behave as self-organized critical systems with weakly chaotic dynamics (Chen et al. 1990; Bak et al. 1990). And complex communities of species appear to evolve towards persistent and emergent states, the species membership of which is largely determined by happenstance (Pimm 1992). A fascinating challenge for the future is to explore the extent to which notions of self-organized criticality apply to the biotic–abiotic interactions which create geoecosystems.

BIBLIOGRAPHY

Aandahl, A. R. (1948) 'The characterization of slope positions and their influence on the total nitrogen content of a few virgin soils of western Iowa', *Soil Science Society of America Proceedings* 13: 449–454.

Abbott, I., Abbott, L. K., and Grant, P. R. (1977) 'Comparative ecology of Galápagos ground finches (*Geospiza* Gould): evaluation of the importance of floristic diversity and interspecific competition', *Ecological Monographs* 47: 151–184.

Adams, A. B., Dale, V. H., Kruckeberg, A. R., and Smith, E. (1987) 'Plant survival, growth form and regeneration following the May 18, 1980, eruption of Mount St. Helens, Washington', *Northwest Science* 61: 160–170.

Ager, D. V. (1993) *The New Catastrophism: The Importance of the Rare Event in Geological History*, Cambridge: Cambridge University Press.

Ahlgren, I. F. and Ahlgren, C. E. (1960) 'Ecological effects of forest fires', *The Botanical Review* 26: 483–533.

Ahrens, T. J. and Harris, A. W. (1992) 'Deflection and fragmentation of near-Earth asteroids', *Nature* 360: 429–433.

Aldiss, B. W. (1982) *Helliconia Spring*, London: Jonathan Cape.

Aldiss, B. W. (1983) *Helliconia Summer*, London: Jonathan Cape.

Aldiss, B. W. (1985) *Helliconia Winter*, London: Jonathan Cape.

Aldrich, J. W. and James, F. C. (1991) 'Ecogeographic variation in the American robin, *Turdus migratorius*', *Auk* 108: 230–249.

Alexander, E. B., Mallory, J. I., and Colwell, W. L. (1993) 'Soil–elevation relationships on a volcanic plateau in the Southern Cascade Range, northern California, USA', *Catena* 20: 113–128.

Allee, W. C., Emerson, A. E., Park, O., Park, T., and Schmidt, K. P. (1949) *Principles of Animal Ecology*, Philadelphia: W. B. Saunders.

Allee, W. C. and Park, T. (1939) 'Concerning ecological principles', *Science* 89: 166–169.

Allen, J. A. (1877) 'The influence of physical conditions in the genesis of species', *Radical Review* 1: 108–140.

Allen, J. C., Schaffer, W. M., and Rosko, D. (1993) 'Chaos reduces species extinction by amplifying local population noise', *Nature* 364: 229–232.

Allen, T. F. H. and Hoekstra, T. W. (1992) *Toward a Unified Ecology*, New York: Columbia University Press.

Allen, T. F. H. and Starr, T. B. (1982) *Hierarchy: Perspectives for Ecological Complexity*, Chicago: The University of Chicago Press.

Alvarez, W. (1986) 'Towards a theory of impact crises', *Eos* 67: 649, 653–655.

Alvarez, L. W., Alvarez, W., Asaro, F., and Michel, H. V. (1980) 'Extraterrestrial cause for Cretaceous–Tertiary extinction', *Science* 208: 1095–1108.

Alvarez, L. W., Alvarez, W., Asaro, F., and Michel, H. V. (1982) 'Current status of the impact theory for the terminal Cretaceous extinction', in L. T. Silver and P. H. Schultz (eds) *Geological Implications of Impacts of Large Asteroids and Comets on the Earth*, pp. 306–316, Geological Society of America Special Paper 190.

Amundson, R. and Jenny, H. (1991) 'The place of humans in the state factor theory of ecosystems and their soils', *Soil Science* 151: 99–109.

Anderson, K. E. and Furley, P. A. (1975) 'An assessment of the relationships between surface properties of chalk soils and slope form using principal components analysis', *Journal of Soil Science* 26: 130–143.

Anderson, P. and Shimwell, D. W. (1981) *Wild Flowers and other Plants of the Peak District: An Ecological Study*, Ashbourne, Derbyshire: Moorland Publishing.

Andrewartha, H. G. and Birch, L. C. (1954) *The Distribution and Abundance of Animals*, Chicago: University of Chicago Press.

Archer, M. (1981) 'A review of the origins and radiations of Australian mammals', in A. Keast (ed.) *Ecological Biogeography of Australia, Volume 3* (Monographiae Biologicae, vol. 41), pp. 1436–1488, The Hague: W. Junk.

Arkley, R. J. (1963) 'Calculation of carbonate and water movement in soil from climatic data', *Soil Science* 96: 239–248.

Arkley, R. J. (1967) 'Climates of some Great Soil Groups of the western United States', *Soil Science* 103: 389–400.

Armand, A. D. (1992) 'Sharp and gradual mountain timberlines as a result of species interaction', in A. J. Hansen and F. di Castri (eds) *Landscape Boundaries: Consequences for Biotic Diversity and Ecological Flows* (Ecological Studies, vol. 92), pp. 360–378, New York: Springer.

Arndt, J. L. and Richardson, J. L. (1989) 'Geochemistry of hydric soil salinity in a recharge–throughflow–discharge prairie–pothole wetland system', *Soil Science Society of America Journal* 53: 848–855.

Arndt, J. L. and Richardson, J. L. (1993) 'Temporal variations in the salinity of shallow groundwater from the periphery of some North Dakota wetlands (USA)', *Journal of Hydrology* 141: 75–101.

Arnett, R. R. and Conacher, A. J. (1973) 'Drainage basin expansion and the nine unit landsurface model', *Australian Geographer* 12: 237–249.

Arrhenius, O. (1921) 'Species and area', *Journal of Ecology* 9: 95–99.

Asher, D. J. and Clube, S. V. M. (1993) 'An extraterrestrial influence during the current glacial–interglacial', *Quarterly Journal of the Royal Astronomical Society* 34: 481–511.

Auerbach, M. and Shmida, A. (1993) 'Vegetation change along an altitudinal gradient on Mt Hermon, Israel – no evidence for discrete communities', *Journal of Ecology* 81: 25–33.

Aulitsky, H., Turner, H., and Meyer, H. (1982) 'Bioklimatische Grundlagen einer standortsgemässen Bewirtschaftung des subalpinen Lärchen-Arvenwaldes', *Mitteilungen der Eidgenössischen Zentralanstalt für das forstliche Versuchswesen* 58: 325–580.

Austin, M. P. and Smith, T. M. (1989) 'A new model for the continuum concept', *Vegetatio* 83: 35–47.

Bach, R. (1950) 'Die Standorte jurassischer Buchenwaldgesellschaften mit besonderer Berücksichtigung der Böden (Humuskarbonatböden und Rendzinen)', *Berichte der Schweizerischen Botanischen Gesellschaft* 60: 51–162.

Bagnouls, F. and Gaussen, H. (1953) 'Période de sécheresse et végétation', *Comptes rendus hebdomadaires des Seances de l'Académie des Sciences, Paris* 236: 1075–1077.

Bagnouls, F. and Gaussen, H. (1957) 'Les climats biologiques et leurs classification', *Annales de Géographie* 355: 193–220.

Bailey, I. W. and Sinnott, E. W. (1916) 'The climatic distribution of certain types of angiosperm leaves', *American Journal of Botany* 3: 24–39.

Bailey, M. E., Clube, S. V. M., and Napier, W. M. (1990) *The Origin of Comets*, Oxford: Pergamon Press.

Bailey, M. E., Clube, S. V. M., Hahn, G., Napier, W. M., and Valsecchi, G. B. (1994) 'Hazards due to giant comets: climate and short-term catastrophism', in T. Gehrels, M. S. Matthews, and A. M. Schumann (eds) *Hazards Due To Comets and Asteroids*, in press, Tucson, Arizona: University of Arizona Press.

Baillie, M. G. L. and Munro, M. A. R. (1988) 'Irish tree rings, Santorini and volcanic dust veils', *Nature* 332: 344–346.

Baize, D. (1993) 'Place of horizons in the new French "Référentiel Pédologique" ', *Catena* 20: 383–394.

Bak, P. and Chen, K. (1991) 'Self-organized criticality', *Scientific American* 264: 46–53.

Bak, P., Chen, K., and Creutz, M. (1989) 'Self-organized criticality and the "game of life" ', *Nature* 342: 780–782.

Bak, P., Chen, K., and Tang, C. (1990) 'A forest-fire model and some thoughts on turbulence', *Physics Letters A* 147: 297–300.

Bak, P., Tang, C., and Wiesenfeld, K. (1988) 'Self-organized criticality', *Physical Review A* 38: 364–374.

Baker, A. J. M., Proctor, J., and Reeves, R. D. (eds) (1992) *The Vegetation of Ultramafic (Serpentine) Soils*, Andover, Hants: Intercept.

Ballantyne, A. K. (1963) 'Recent accumulation of salts in the soils of south-eastern Saskatchewan', *Canadian Journal of Soil Science* 43: 52–58.

Balling, R. C., Jr, Meyer, G. A., and Wells, S. G. (1992) 'Climate change in Yellowstone National Park: is the drought-related risk of wildfires increasing?', *Climatic Change* 22: 35–45.

Barrett, G. W., Van Dyne, G. M., and Odum, E. P. (1976) 'Stress ecology', *BioScience* 26: 192–194.

Barry, R. G. (1992) *Mountain Weather and Climate*, 2nd edn, London and New York: Routledge.

Barshad, I. (1958) 'Factors affecting soil formation', *Clay and Clay Minerals* 6: 110–132.

Bartell, S. M. and Brenkert, A. L. (1991) 'A spatial–temporal model of nitrogen dynamics in a deciduous forest watershed', in M. G. Turner and R. H. Gardner (eds) *Quantitative Methods in Landscape Ecology: the Analysis and Interpretation of Landscape Heterogeneity*, pp. 379–398, New York: Springer.

Barton, A. M. (1993) 'Factors controlling plant distributions: drought, competition, and fire in montane pines in Arizona', *Ecological Monographs* 63: 367–397.

Bates, R. L. and Jackson, J. A. (1980) *Glossary of Geology*, 2nd edn, Falls Church, Virginia: American Geological Institute.

Baumgardner, G. D. and Kennedy, M. L. (1993) 'Morphometric variation in kangaroo rats (Genus *Dipodomys*) and its relationship to selected abiotic variables', *Journal of Mammalogy* 74: 69–85.

Baumgartner, A. (1980) 'Mountain climates from a perspective of forest growth', in U. Benecke and M. R. Davis (eds) *Mountain Environments and Sub-Alpine Tree Growth*, pp. 27–39, New Zealand Forest Service, Technical Paper No. 70.

Bäumler, R. and Zech, W. (1994) 'Soils of the high mountain region of Eastern Nepal: classification, distribution and soil forming processes', *Catena* 22: 85–103.

Bellamy, D. (1976) *Bellamy's Europe*, London: British Broadcasting Corporation.

Bennett, K. D. (1993) 'Holocene forest dynamics with respect to southern Ontario', *Review of Palaeobotany and Palynology* 79: 69–81.

Bergmann, C. (1847) 'Über die Verhältnisse der Wärmeökonomie der Thiere zu ihrer Grösse', *Göttinger Studien* 3: 595–708.

Bertalanffy, L. von (1950) 'The theory of open systems in physics and biology', *Science* 111: 23–29.

Billings, W. D. (1950) 'Vegetation and plant growth as affected by chemically altered rocks in the western Great Basin', *Ecology* 31: 62–74.

Billings, W. D. (1952) 'The environmental complex in relation to plant growth and distribution', *The Quarterly Review of Biology* 27: 251–265.

Billings, W. D. (1954) 'Temperature inversions in the piñon–juniper zone of a Nevada mountain range', *Butler University Botanical Studies* 11: 112–118.

Billings, W. D. (1990) 'The mountain forests of North America and their environments', in C. B. Osmond, L. F. Pitelka, and G. M. Hidy (eds) *Plant Biology of the Basin and Range* (Ecological Studies, vol. 80), pp. 47–86, Berlin: Springer.

Birkeland, P. W. (1984) *Soils and Geomorphology*, New York and Oxford: Oxford University Press.

Birkeland, P. W., Burke, R. M., and Benedict, J. B. (1989) 'Pedogenic gradients for iron and aluminium accumulation and phosphorus depletion in Arctic and Alpine soils as a function of time and climate', *Quaternary Research* 32: 193–204.

Blackman, F. F. (1905) 'Optima and limiting factors', *Annals of Botany* 19: 281–295.

Blume, H.-P. (1968) 'Die pedogenetische Deutung einer Catena durch die Untersuchung der Bodendynamik', *Transactions of the Ninth International Congress of Soil Science, Adelaide* 4: 441–449.

Blume, H.-P. and Schlichting, E. (1965) 'The relationships between historical and experimental pedology', in E. G. Hallsworth and D. V. Crawford (eds) *Experimental Pedology*, pp. 340–353, London: Butterworths.

Bolin, B. (1980) *Climatic Changes and their Effects on the Biosphere*, Geneva, Switzerland: World Meteorological Organization, Publication No. 542.

Bonan, G. B., Pollard, D., and Thompson, S. L. (1992) 'Effects of boreal forest vegetation on global climate', *Nature* 359: 716–718.

Borcard, D., Legendre, P., and Drapeau, P. (1992) 'Partialling out the spatial component of ecological variation', *Ecology* 73: 1045–1055.

Borhidi, A. (1991) *Phytogeography and Vegetation Ecology of Cuba*, Budapest: Akadémiai Kiadó.

Botkin, D. B. (1990) *Discordant Harmonies: A New Ecology for the Twenty-First Century*, New York: Oxford University Press.

Botkin, D. B. (1993) *Forest Dynamics: An Ecological Model*, Oxford and New York: Oxford University Press.

Box, E. O. (1981) *Macroclimate and Plant Forms: An Introduction to Predictive Modelling in Phytogeography*, The Hague: W. Junk.

Box, E. O. and Meentemeyer, V. (1991) 'Geographic modeling and modern ecology', in G. Esser and D. Overdieck (eds) *Modern Ecology: Basic and Applied Aspects*, pp. 773–804, Amsterdam: Elsevier.

Box, E. O., Crumpacker, D. W., and Hardin, E. D. (1993) 'A climatic model for location of plant species in Florida, USA', *Journal of Biogeography* 20: 629–644.

Boyce, M. S. (1978) 'Climatic variability and body size variation in the muskrats (*Ondatra zibethicus*) of North America', *Oecologia* 36: 1–19.

Brewer, R., Crook, A. W., and Speight, J. A. (1970) 'Proposal for soil-stratigraphic units in the Australian stratigraphic code', *Journal of the Geological Society of Australia* 17: 103–111.

Bridges, E. M. (1978) *World Soils*, 2nd edn, Cambridge: Cambridge University Press.

Bridges, E. M. (1993) 'Soil horizon designations; past use and future prospects', *Catena* 20: 363–373.

Brokaw, N. V. L. (1985) 'Treefalls, regrowth, and community structure in tropical forests', in S. T. A. Pickett and P. S. White (eds) *The Ecology of Natural Disturbance and Patch Dynamics*, pp. 53–69, Orlando, Florida: Academic Press.

Brooks, R. R. (1987) *Serpentine and Its Vegetation: A Multidisciplinary Approach*, London and Sydney: Croom Helm.

Brown, J. H. (1971) 'Mammals on mountaintops: nonequilibrium insular biogeography', *American Naturalist* 105: 467–478.

Brown, J. H. (1986) 'Two decades of interaction between the MacArthur–Wilson model and the complexities of mammalian distributions', *Biological Journal of the Linnean Society* 28: 231–251.

Brown, J. H. and Kodric-Brown, A. (1977) 'Turnover rates in insular biogeography: effect of immigration on extinction', *Ecology* 58: 445–449.

Brown, J. H. and Maurer, B. A. (1989) 'Macroecology: the division of foods and space among species on continents', *Science* 243: 1145–1150.

Brown, R. P. and Thorpe, R. S. (1991a) 'Within-island microgeographic variation in body dimensions and scalation of the skink *Chalcides sexlineatus*, with testing of causal hypotheses', *Biological Journal of the Linnean Society* 44: 47–64.

Brown, R. P. and Thorpe, R. S. (1991b) 'Within-island microgeographic variation in the colour pattern of the skink *Chalcides sexlineatus*: pattern and cause', *Journal of Evolutionary Biology* 4: 557–574.

Brown, R. P., Thorpe, R. S., and Báez, M. (1991) 'Parallel within-islands micro-evolution of lizards on neighbouring islands', *Nature* 352: 60–62.

Browne, J. (1983) *The Secular Ark: Studies in the History of Biogeography*, New Haven, Connecticut: Yale University Press.

Brues, C. T. (1946) *Insect Dietary*, Cambridge, Massachusetts: Harvard University Press.

Bryson, R. A. (1966) 'Air masses, streamlines, and the boreal forest', *Geographical Bulletin* 8: 228–269.

Büdel, J. (1982) *Climatic Geomorphology*, translated by Lenore Fischer and Detlef Busche, Princeton, New Jersey: Princeton University Press.

Budyko, M. I. (1974) *Climate and Life* (International Geophysics Series, Vol. 18), English edn ed. David H. Miller, New York: Academic Press.

Bunting, B. T. (1965) *The Geography of Soil*, London: Hutchinson.

Buol, S. W., Hole, F. D., and McCracken, R. J. (1980) *Soil Genesis and Classification*, 2nd edn, Ames, Iowa: The Iowa State University Press.

Burke, I. C., Yonker, C. M., Parton, W. J., Cole, C. V., Flach, K., and Schimel, D. S. (1989) 'Texture, climate, and cultivation effects on soil organic matter content in U.S. grassland soils', *Journal of the Soil Science Society of America* 53: 800–805.

Bushnell, T. M. (1942) 'Some aspects of the soil catena concept', *Soil Science Society of America Proceedings* 7: 466–476.

Bushnell, T. M. (1945) 'The catena cauldron', *Soil Science Society of America Proceedings* 10: 335–340.

Butcher, S. S., Orians, G. H., Charlson, R. J., and Wolfe, G. V. (eds) (1992) *Global Biogeochemical Cycles*, New York and London: Academic Press.

Butler, B. E. (1982) 'A new system for soil studies', *Journal of Soil Science* 33: 581–595.

Butler, D. R. (1992) 'The grizzly bear as an erosional agent in mountainous terrain', *Zeitschrift für Geomorphologie* NF 36: 179–189.

Cameron, G. N. and McClure, P. A. (1988) 'Geographic variation in life history traits of the hispid cotton rat (*Sigmodon hispidus*), in M. S. Boyce (ed.) *Evolution and*

Life Histories of Mammals: Theory and Pattern, pp. 33–64, New Haven, Connecticut: Yale University Press.

Cantlon, J. E. (1953) 'Vegetation and microclimates on north and south slopes of Cushetunk Mountain, New Jersey', *Ecological Monographs* 23: 241–270.

Carlquist, S. (1965) *Island Life: A Natural History of the Islands of the World*, Garden City, New York: Published for the American Museum of Natural History by the Natural History Press.

Carter, B. J. and Ciolkosz, E. J. (1991) 'Slope gradient and aspect effects on soils developed from sandstone in Pennsylvania', *Geoderma* 49: 199–213.

Catt, J. A. (1988) 'Soils of the Plio-Pleistocene: do they distinguish types of interglacial?', *Philosophical Transactions of the Royal Society of London* 318B: 539–557.

Cattelino, P. J., Noble, I. R., Slayter, R. O., and Kessell, S. R. (1979) 'Predicting the multiple pathways of plant succession', *Environmental Management* 3: 41–50.

Chartres, C. J. and Pain, C. F. (1984) 'A climosequence of soils on Late Quaternary volcanic ash in highland Papua New Guinea', *Geoderma* 32: 131–155.

Chen, K., Bak, P., and Jensen, M. H. (1990) 'A deterministic critical forest fire model', *Physics Letters A* 149: 207–210.

Chew, R. M. (1974) 'Consumers as regulators of ecosystems: an alternative view to energetics', *Ohio Journal of Science* 74: 359–370.

Christensen, N. L., Agee, J. K., Brussard, P. F., Hughes, J., Knight, D. H., Minshall, G. W., Peek, J. M., Pyne, S. J., Swanson, F. J., Thomas, J. W., Wells, S., Williams, S. E., and Wright, H. A. (1989) 'Interpreting the Yellowstone fires of 1988', *BioScience* 39: 678–685.

Chyba, C. F., Thomas, P. J., and Zahnle, K. J. (1993) 'The 1908 Tunguska explosion: atmospheric disruption of a stony asteroid', *Nature* 361: 40–44.

Clayden, B. (1982) 'Soil classification', in E. M. Bridges and D. A. Davidson (eds) *Principles and Applications of Soil Geography*, pp. 58–96, London and New York: Longman.

Clements F. E. (1916) *Plant Succession: An Analysis of the Development of Vegetation*, Washington: Carnegie Institute, Publication No. 242.

Clements, F. E. (1936) 'Nature and structure of the climax', *Journal of Ecology* 24: 252–284.

Clements, F. E. and Shelford, V. E. (1939) *Bio-Ecology*, New York: John Wiley & Sons.

Cloudsley-Thompson, J. L. (1975a) *Terrestrial Environments*, London: Croom Helm.

Cloudsley-Thompson, J. L. (1975b) *The Ecology of Oases*, Watford, Hertfordshire: Merrow.

Clube, S. V. M. and Napier, W. M. (1990) *The Cosmic Winter*, Oxford: Basil Blackwell.

Coffin, M. F. and Eldholm, O. (1993) 'Large igneous provinces', *Scientific American* 269: 26–33.

Cole, L. C. (1958) 'The ecosphere', *Scientific American* 198: 83–96.

Cole, M. M. (1986) *The Savannas: Biogeography and Geobotany*, London: Academic Press.

Commoner, B. (1972) *The Closing Circle: Confronting the Environmental Crisis*, London: Jonathan Cape.

Conacher, A. J. (1975) 'Throughflow as a mechanism responsible for excessive soil salinization in non-irrigated, previously arable lands in the western Australian wheatbelt: field study', *Catena* 2: 31–68.

Conacher, A. J. and Dalrymple, J. B. (1977) 'The nine-unit landsurface model: an approach to pedogeomorphic research', *Geoderma* 18: 1–154.

Cooper, A. W. (1960) 'An example of the role of microclimate in soil genesis', *Soil Science* 86:109–120.

Cooper, C. F. (1960) 'Changes in vegetation, structure and growth of southwestern pine forests since white settlement', *Ecological Monographs* 30: 129–164.

Cosby, B. J., Wright, R. F., Hornberger, G. M., and Galloway, J. N. (1985) 'Modelling the effects of acid deposition: estimation of long-term water quality responses in a small forested catchment', *Water Resources Research* 21: 1591–1601.

Courtillot, V. (1990) 'Deccan volcanism at the Cretaceous–Tertiary boundary: past climatic crises as a key to the future?', *Palaeogeography, Palaeoclimatology, Palaeoecology (Global and Planetary Change Section)* 89: 291–299.

Courtillot, V. and Cisowski, S. (1987) 'The Cretaceous/Tertiary boundary events: external or internal causes?', *Eos* 68: 193, 200.

Courtillot, V., Féraud, G., Maluski, H., Vandamme, D., Moreau, M. G., and Besse, J. (1988) 'Deccan flood basalts and the Cretaceous/Tertiary boundary', *Nature* 333: 843–846.

Cousins, S. H. (1989) Letter under 'Species richness and the energy theory', *Nature* 340: 350–351.

Covey, C., Ghan, S. J., Walton, J. J., and Weissman, P. R. (1990) 'Global environmental effects of impact-generated aerosols; results from a general circulation mode', in V. L. Sharpton and P. D. Ward (eds) *Global Catastrophes in Earth History; An Interdisciplinary Conference on Impacts, Volcanism, and Mass Mortality*, pp. 263–270, Geological Society of America Special Paper 247.

Cox, K. G. (1988) 'Gradual volcanic catastrophes?', *Nature* 333: 802.

Crawley, M. J. (1983) *Herbivory: the Dynamics of Animal–Plant Interactions* (Studies in Ecology, vol. 10), Oxford: Blackwell Scientific Publications.

Crocker, R. L. (1952) 'Soil genesis and the pedogenic factors', *The Quarterly Review of Biology* 27: 139–168.

Crook, I. G. (1975) 'The tuatara', in G. Kuschel (ed.) *Biogeography and Ecology in New Zealand* (Monographiae Biologicae, vol. 27), pp. 331–352, The Hague: W. Junk.

Crowell. K. L. (1986) 'A comparison of relict versus equilibrium models for insular mammals of the Gulf of Maine', *Biological Journal of the Linnean Society* 28: 37–64.

Culling, W. E. H. (1988) 'A new view of the landscape', *Transactions of the Institute of British Geographers*, New Series 13: 345–360.

Currie, D. J. (1991) 'Energy and large-scale patterns of animal and plant species richness, *American Naturalist* 137: 27–49.

Currie, D. J. and Fritz, J. T. (1993) 'Global patterns of animal abundance and species energy use', *Oikos* 67: 56–68.

Currie, D. J. and Paquin, V. (1987) 'Large-scale biogeographical patterns of species richness of trees', *Nature* 329: 326–327.

Dahlgren, R. A. (1994) 'Soil acidification and nitrogen saturation from weathering of ammonium-bearing rock', *Nature* 368: 838–841.

Damuth, J. (1993) 'Cope's rule, the island rule and the scaling of mammalian population density', *Nature* 365: 748–750.

Dan, J. and Yaalon, D. H. (1968) 'Pedomorphic forms and pedomorphic surfaces', *Transactions of the Ninth International Congress of Soil Science, Adelaide* 4: 577–584.

Daniels, R. B. and Hammer, R. D. (1992) *Soil Geomorphology*, New York: John Wiley & Sons.

Daniels, R. J. R. (1992) 'Geographical distribution patterns of amphibians in the Western Ghats, India', *Journal of Biogeography* 19: 521–529.

Dansereau, P. (1957) *Biogeography: An Ecological Perspective*, New York: Ronald Press.

Darlington, P. J., Jr (1957) *Zoogeography: the Geographical Distribution of Animals*, New York: John Wiley & Sons.

Darwin, C. R. (1859) *The Origin of Species by means of Natural Selection, or the Preservation of Favoured Races in the Struggle for Life*, London: John Murray.

Dasmann, R. F. (1976) *Environmental Conservation*, 4th edn, New York: John Wiley & Sons.

Daubenmire, R. (1954) 'Alpine timberlines in the Americas and their interpretation', *Butler University Botanical Studies* 11: 119–136.

de Blij, H. J. and Muller, P. O. (1993) *Physical Geography of the Global Environment*, New York: John Wiley & Sons.

de Laubenfels, M. W. (1956) 'Dinosaur extinction: one more hypothesis', *Journal of Paleontology* 30: 207–212.

Delcourt, H. R. and Delcourt, P. A. (1988) 'Quaternary landscape ecology: relevant scales in space and time', *Landscape Ecology* 2: 23–44.

DeLucia, E. H. and Schlesinger, W. H. (1990) 'Ecophysiology of Great Basin and Sierra Nevada vegetation on contrasting soils', in C. B. Osmond, L. F. Pitelka, and G. M. Hidy (eds) *Plant Biology of the Basin and Range* (Ecological Studies, vol. 80), pp. 143–178, Berlin: Springer.

DeLucia, E. H., Schlesinger, W. H., and Billings, W. D. (1989) 'Edaphic limitations to growth and photosynthesis in Sierran and Great Basin vegetation', *Oecologia* 78: 184–190.

DeRose, R. C., Trustrum, N. A., and Blaschke, P. M. (1991) 'Geomorphic change implied by regolith–slope relationships on steepland hillslopes, Taranaki, New Zealand', *Catena* 18: 489–514.

Diamond, J. M. (1974) 'Colonization of exploded volcanic islands by birds: the supertramp strategy', *Science* 184: 803–806.

Dikau, R. (1989) 'The application of a digital relief model to landform analysis in geomorphology', in J. Raper (ed.) *Three-dimensional Applications in Geographic Information Systems*, pp. 51–77, London and New York: Taylor & Francis.

Docters van Leeuwen, W. M. (1936) 'Krakatau, 1883 to 1933', *Annales du Jardin Botanique de Buitenzorg* 56–57: 1–506.

Doyle, T. W. (1981) 'The role of disturbance in the gap dynamics of a montane rain forest: an application of a tropical forest succession model', in D. C. West, H. H. Shugart, and D. B. Botkin (eds) *Forest Succession: Concepts and Applications*, pp. 56–73, New York: Springer.

Duchaufour, P. (1982) *Pedology: Pedogenesis and Classification*, translated by T. R. Paton, London: George Allen & Unwin.

Duncan, R. A. and Pyle, D. G. (1988) 'Rapid eruption of the Deccan flood basalts at the Cretaceous/Tertiary boundary', *Nature* 333: 841–843.

Durgin, P. (1984) 'Subsurface drainage erodes forested granitic terrane', *Physical Geography* 4: 24–39.

Duvigneaud, P. (1974) *La Synthèse écologique: Populations, Communautés, Écosystèmes, Biosphère, Noosphère*, Paris: Doin.

Ehleringer, J. R., Mooney, H. A., Rundel, P. W., Evans, R. D., Palma, B., and Delatorre, J. (1992) 'Lack of nitrogen cycling in the Atacama Desert', *Nature* 359: 316–318.

Eldredge, N. (1985) *Unfinished Synthesis: Biological Hierarchies and Modern Evolutionary Thought*, New York and Oxford: Oxford University Press.

Elkins, N. (1989) Letter under 'Species richness and the energy theory', *Nature* 340: 350.

Elton, C. S. (1958) *The Ecology of Invasions by Animals and Plants*, London: Chapman and Hall.

Emiliani, C., Kraus, E. B., and Shoemaker, E. M. (1981) 'Sudden death at the end of the Mesozoic', *Earth and Planetary Science Letters* 55: 317–334.

Endler, J. A. (1977) *Geographic Variation, Speciation, and Clines* (Monographs in Population Biology 10), Princeton, New Jersey: Princeton University Press.

Esser, G. and Lieth, H. (1989) 'Decomposition in tropical rain forests compared with other parts of the world', in H. Lieth and M. J. A. Werger (eds) *Tropical Rain Forest Ecosystems*, pp. 571–580, Amsterdam: Elsevier.

Evans, L. J. (1978) 'Quantification and pedological processes', in W. C. Mahaney (ed.) *Quaternary Soils*, pp. 361–378, Norwich: Geo Abstracts.

Faeth, S. H. and Kane, T. C. (1978) 'Urban biogeography: city parks as islands for Diptera and Coleoptera', *Oecologia* 32: 127–133.

Fenneman, N. M. (1916) 'Physiographic divisions of the United States', *Annals of the Association of American Geographers* 6: 19–98.

Finney, H. R., Holowaychuk, N., and Heddleson, M. R. (1962) 'The influence of microclimate on the morphology of certain soils of the Allegheny Plateau of Ohio', *Soil Science Society of America Proceedings* 26: 287–292.

Flessa, K. W. (1975) 'Area, continental drift and mammalian diversity', *Paleobiology* 1: 189–194.

Flessa, K. W. (1981) 'The regulation of mammalian faunal similarity among continents', *Journal of Biogeography* 8: 427–438.

Florkin, M. (1943) *Introduction à la Biochemie générale*, Liége: Éditions Desoer; Paris: Masson et Cie.

Forman, R. T. T. and Godron, M. (1986) *Landscape Ecology*, New York: John Wiley & Sons.

Forster, J. R. (1778) *Observations made during a Voyage round the World, on Physical Geography, Natural History, and Ethnic Philosophy. Especially on: 1. The Earth and Its Strata; 2. Water and the Ocean; 3. The Atmosphere; 4. The Changes of the Globe; 5. Organic bodies; and 6. The Human Species*, London: printed for G. Robinson.

Fortescue, J. A. C. (1980) *Environmental Geochemistry: A Holistic Approach* (Ecological Studies, vol. 35), New York: Springer.

Foster, D. R. (1988a) 'Disturbance history, community organization and vegetation of the old-growth Pisgah Forest, southwestern New Hampshire, USA', *Journal of Ecology* 76: 105–134.

Foster, D. R. (1988b) Species and stand response to catastrophic wind in central New England, USA, *Journal of Ecology* 76: 135–151.

Foster, J. B. (1964) 'Evolution of mammals on islands', *Nature* 202: 234–235.

Francis, P. (1993) *Volcanoes: A Planetary Perspective*, Oxford: Clarendon Press.

Franklin, B. (1784) 'Meteorological imaginations and conjectures', *Memoirs of the Literary and Philosophical Society of Manchester* 2: 373–377.

Friedman, H. (1985) 'The science of global change – an overview', in T. F. Malone and J. G. Roederer (eds) *Global Change*, pp. 20–52, Cambridge: published on behalf of the ICUS Press by Cambridge University Press.

Furley, P. A. (1968) 'Soil formation and slope development – II. The relationship between soil formation and gradient in the Oxford area', *Zeitschrift für Geomorphologie* NF 12: 25–42.

Furley, P. A. (1971) 'Relationship between slope form and soil properties over chalk parent materials', in D. Brunsden (ed.) *Slopes: Form and Process* (Institute of British Geographers, Special Publication No. 3), pp. 141–163, Kensington Gore, London: Institute of British Geographers.

Gams, H. (1931) 'Die klimatische Begrenzung von Pflanzenarealen und die Vertei-lung der hygrischen Kontinentalität in den Alpen', *Zeitschrift der Gesellschaft für Erdkunde zu Berlin* 1931: 321–346.

Gardner, R. H., O'Neill, R. V., Turner, M. G., and Dale, V. H. (1989) 'Quantifying scale-dependent effects of animal movement with simple percolation models', *Landscape Ecology* 3: 217–227.

Gardner, R. H., Turner, M. G., Dale, V. H., and O'Neill, R. V. (1992) 'A percolation model of ecological flows', in A. J. Hansen and F. di Castri (eds) *Landscape Boundaries: Consequences for Biotic Diversity and Ecological Flows* (Ecological Studies, vol. 92), pp. 259–269, New York: Springer.

Gates, D. M. (1993) *Climate Change and Its Biological Consequences*, Sunderland, Massachusetts: Sinauer Associates.

Gaussen, H. (1963) 'Les cartes bioclimatiques et de la végétation. Principes directeurs et emploi de la couleur', *Science du Sol* (May): 1–14.

Gaussen, H. and Bagnouls, F. (1952) 'L'indice xéromorphique', *Bulletin de l'Associa-tion de Géographes français* (1952): 222–223.

Gerrard, A. J. (1981) *Soils and Landforms: An Integration of Geomorphology and Pedology*, London: George Allen and Unwin.

Gerrard, A. J. (1988) *Soil–Slope Relationships: A Dartmoor Example*, University of Birmingham, Occasional Paper No. 26, School of Geography, University of Birmingham.

Gerrard, A. J. (1990a) 'Soil variations on hillslopes in humid temperate climates', *Geomorphology* 3: 225–244.

Gerrard, A. J. (1990b) *Mountain Environments: An Examination of the Physical Geography of Mountains*, London: Belhaven Press.

Gerrard, A. J. (1991) 'The status of temperate hillslopes in the Holocene', *The Holocene* 1: 86–90.

Gerrard, A. J. (1992) *Soil Geomorphology: An Integration of Pedology and Geomor-phology*, London: Chapman and Hall.

Gerrard, A. J. (1993) 'Soil geomorphology – present dilemmas and future challenges', *Geomorphology* 7: 61–84.

Gibbs, H. S. (1980) *New Zealand Soils: An Introduction*, Wellington: Oxford University Press.

Gilbert, F. S. (1980) 'The equilibrium theory of island biogeography: fact or fiction?', *Journal of Biogeography* 7: 209–235.

Gillard, A. (1969) 'On terminology of biosphere and ecosphere', *Nature* 223: 500–501.

Gilpin, M. E. and Diamond, J. M. (1976) 'Calculation of immigration and extinction curves from the species–area–distance relation', *Proceedings of the National Academy of Sciences of the United States of America, Washington DC* 73: 4130–4134.

Givnish, T. J. (1987) 'Comparative studies of leaf form: assessing the relative roles of selective pressures and phylogenetic constraints', *New Phytologist* 106 (Supple-ment): 131–160.

Glazovskaya, M. A. (1963) 'On geochemical principles of the classification of natural landscapes', *International Geology Review* 5: 1403–1431.

Glazovskaya, M. A. (1968) 'Geochemical landscapes and geochemical soil sequences', *Transactions of the Ninth International Congress of Soil Science, Adelaide* 4: 303–312.

Gleason, H. A. (1922) 'On the relation between species and area', *Ecology* 3: 158–162.

Gleason, H. A. (1925) 'Species and area', *Ecology* 6: 66–74.

Gleason, H. A. (1926) 'The individualistic concept of the plant association', *Bulletin of the Torrey Botanical Club* 53: 7–26.

Glinka, K. D. (1914) *Die Typen der Bodenbildung, ihre Klassifikation und geographische Verbreitung*, Berlin: Gebrüder Borntraeger. Translated by C. F. Marbut, 1927, as *The Great Soil Groups of the World*, Ann Arbor, Michigan: Edwards Bros.

Gloger, C. W. L. (1833) *Das Abändern der Vögel durch Einfluss des Klimas: Nach zoologischen, zunächst von dem europäischen Landvälgeln, entnommenen Beobachtungen dargestellt, mit den entsprechenden Erfahrungen bei den europäischen Säugthieren verglichen, und durch Tatsachen aus dem Gebiete der Physiologie, der Physik und der physischen Geographie erläutert*, Breslau: A. Schultz.

Golley, F. B. (1978) 'Series editor's foreword', in H. F. H. Leith and R. H. Whittaker (eds) *Patterns of Primary Production in the Biosphere* (Benchmark Papers in Ecology, vol. 8), pp. v–vi, Stroudsberg, Pennsylvania: Dowden, Hutchinson & Ross.

Gorman, M. L. (1979) *Island Ecology* (Outline Studies in Ecology), London: Chapman and Hall.

Grace, J. (1987) 'Climatic tolerance and the distribution of plants', *New Phytologist* 106 (Supplement): 113–130.

Graham, R. W. and Mead, J. I. (1987) 'Environmental fluctuations and evolution of mammal faunas during the last deglaciation', in W. F. Ruddiman and H. E. Wright Jr (eds) *North America and Adjacent Oceans During the Last Deglaciation* (The Geology of North America, vol. K-3), pp. 371–402, Boulder, Colorado: The Geological Society of America.

Green, D. G. (1989) 'Simulated effects of fire, dispersal and spatial pattern on competition with forest mosaics', *Vegetatio* 82: 139–153.

Grieve, I. C., Proctor, J., and Cousins, S. A. (1990) 'Soil variation with altitude on Volcán Barva, Costa Rica', *Catena* 17: 525–534.

Grime, J. P. (1989) 'The stress debate: symptom of impending synthesis?', *Biological Journal of the Linnean Society* 37: 3–17.

Grinevald, J. (1988) 'Sketch for a history of the idea of the biosphere', in P. Bunyard and E. Goldsmith (eds) *Gaia, the Thesis, the Mechanisms and the Implications*, pp. 1–34, Camelford, Cornwall: Wadebridge Ecological Centre.

Hairston, N. G., Smith, F. E., and Slobodkin, L. B. (1960) 'Community structure, population control and competition', *American Naturalist* 94: 421–425.

Hall, E. R. (1946) *Mammals of Nevada*, Berkeley, California: University of California Press.

Hallsworth, E. G. (1965) 'The relationship between experimental pedology and soil classification', in E. G. Hallsworth and D. V. Crawford (eds) *Experimental Pedology*, pp. 354–374, London: Butterworths.

Hanawalt, R. B. and Whittaker, R. H. (1976) 'Altitudinally coordinated patterns of soils and vegetation in the San Jacinto Mountains, California', *Soil Science* 121: 114–124.

Hanski, I. (1986) 'Population dynamics of shrews on small islands accord with the equilibrium model', *Biological Journal of the Linnean Society* 28: 23–36.

Hanski, I., Turchin, P., Korpimäki, E., and Henttonen, H. (1993) 'Population oscillations of boreal rodents: regulation by mustelid predators leads to chaos', *Nature* 364: 232–235.

Hauhs, M. (1986) 'A model of ion transport through a forested catchment at Lange Bramke, West Germany', *Geoderma* 38: 97–113.

Heaney, L. R. (1978) 'Island area and body size of insular mammals: evidence from the tri-coloured squirrel (*Callosciurus prevosti*) of southeast Asia', *Evolution* 32: 29–44.

Heaney, L. R. (1986) 'Biogeography of mammals in SE Asia: estimates of rates of colonization, extinction and speciation', *Biological Journal of the Linnean Society* 28: 127–165.

Heaney, L. R. and Rickart, E. A. (1990) 'Correlations of clades and clines: geographic, elevational, and phylogenetic distribution patterns among Philippine mammals', in G. Peters and R. Hutterer (eds) *Symposium on Vertebrate Biogeography and Speciation in the Tropics*, pp. 321–332, Bonn: Museum Alexander Koenig.

Heaney, L. R., Heideman, P. D., Rickart, E. A., Utzurrum, R. C. B., and Klompen, J. S. H. (1989) 'Elevational zonation of mammals in the central Philippines', *Journal of Tropical Ecology* 5: 259–280.

Hesse, R., Allee, W. C., and Schmidt, K. P. (1937) *Ecological Animal Geography*, an authorized, rewritten edn based on *Tiergeographie auf ökologische Grundlage* by Richard Hesse, prepared by W. C. Allee and K. P. Schmidt, New York: John Wiley & Sons; London: Chapman and Hall.

Hildebrand, A. R., Penfield, G. T., Kring, D. A., Pilkington, M., Camargo, Z. A., Jacobsen, S. B., and Boynton, W. V. (1991) 'Chicxulub crater: a possible Cretaceous/Tertiary boundary impact crater on the Yucatán Peninsula, Mexico', *Geology* 19: 867–871.

Hilgard, E. W. (1892) *A Report on the Relation of Soil to Climate*, US Department of Agriculture, Weather Bureau, Bulletin No. 3, Washington, DC: US Government Printing Office.

Hills, J. G. (1981) 'Comet showers and the steady-state infall of comets from the Oort cloud', *The Astronomical Journal* 86: 1730–1740.

Hodych, J. P. and Dunning, G. R. (1992) 'Did the Manicouagan impact trigger end-of-Triassic mass extinction?', *Geology* 20: 51–54.

Holdridge, L. R. (1947) 'Determination of world plant formations from simple climatic data', *Science* 105: 367–368.

Holowaychuk, N., Gersper, P. L., and Wilding, L. P. (1969) 'Strontium-90 content of soils near Cape Thompson, Alaska', *Soil Science* 107: 137–144.

Hopkins, D. G., Sweeney, M. D., and Richardson, J. L. (1991) 'Dispersive erosion and Entisol-panpot genesis in sodium-affected landscapes', *Soil Science Society of America Journal* 55: 171–177.

Hubbell, S. P. (1979) 'Tree dispersion, abundance, and diversity in a tropical dry forest', *Science* 203: 1299–1309.

Huggett, R. J. (1973) 'Soil landscape systems: theory and field evidence', Unpublished Ph.D. thesis, University of London.

Huggett, R. J. (1975) 'Soil landscape systems: a model of soil genesis', *Geoderma* 13: 1–22.

Huggett, R. J. (1976) 'Lateral translocation of soil plasma through a small valley basin in the Northaw Great Wood, Hertfordshire', *Earth Surface Processes* 1: 99–109.

Huggett, R. J. (1980) *Systems Analysis in Geography*, Oxford: Clarendon Press.

Huggett, R. J. (1982) 'Models and spatial patterns of soils', in E. M. Bridges and D. A. Davidson (eds) *Principles and Applications of Soil Geography*, pp. 132–170, London and New York: Longman.

Huggett, R. J. (1988) 'Dissipative systems: implications for geomorphology', *Earth Surface Processes and Landforms* 13: 45–49.

Huggett, R. J. (1991) *Climate, Earth Processes and Earth History*, Heidelberg: Springer.

Huggett, R. J. (1993) *Modelling the Human Impact on Nature: Systems Analysis of Environmental Problems*, Oxford: Oxford University Press.

Humboldt, A. von (1817) *De Distributionae Geographica Plantarum*, Paris: Libraria Graeco-Latino-Germanica.

Humboldt, A. von (1849) *Cosmos: A Sketch of a Physical Description of the Universe*, vol. i, translated from the German by E. C. Otté, London: Henry G. Bohn.

Huntley, B., Bartlein, P. J., and Prentice, I. C. (1989) 'Climatic control of the distribution and abundance of beech (*Fagus* L.) in Europe and North America', *Journal of Biogeography* 16: 551–560.

Hustich, I. (1979) 'Ecological concepts and biogeographical zonation in the North: the need for a generally accepted terminology', *Holarctic Ecology* 2: 208–217.

Hutchinson, G. E. (1965) *The Ecological Theater and the Evolutionary Play*, New Haven, Connecticut: Yale University Press.

Hutchinson, G. E. (1970) 'The biosphere', *Scientific American* 223: 45–53.

Huxley, J. S. (1942) *Evolution: The Modern Synthesis*, London: George Allen & Unwin.

Ibáñez, J. J., Ballestra, R. J., and Alvarez, A. G. (1990) 'Soil landscapes and drainage basins in Mediterranean mountain areas', *Catena* 17: 573–583.

Inger, R. F., Shaffer, H. B., Koshy, M., and Badke, R. (1987) 'Ecological structure of a herpetological assemblage in south India', *Amphibia–Reptilia* 8: 189–202.

Iversen, J. (1944) '*Viscum, Hedera* and *Ilex* as climate indicators', *Geologiska föreningens i Stockholm förhandlinger* 66: 463–483.

Ives, J. D. (1992) 'The Andes: geoecology of the Andes', in P. B. Stone (ed.) *The State of the World's Mountains: A Global Report*, pp. 185–256. London and New Jersey: Zed Books.

Jalas, J. and Suominen, J. (1976) *Atlas Florae Europaeae. 3. Salicaceae to Balanophoraceae*, Helsinki: Committee for Mapping the Flora of Europe.

James, F. C. (1970) 'Geographical size variation in birds and its relationship to climate', *Ecology* 51: 365–390.

James, F. C. (1991) 'Complementary descriptive and experimental studies of clinal variation in birds', *American Zoologist* 31: 694–706.

James, J. C., Grace, J., and Hoad, S. P. (1994) 'Growth and photosynthesis of *Pinus sylvestris* at its altitudinal limit in Scotland', *Journal of Ecology* 82: 297–306.

Jarvis, P. G. and McNaughton, K. G. (1986) 'Stomatal control of transpiration: scaling up from leaf to region', *Advances in Ecological Research* 15: 1–49.

Jenny, H. (1930) 'An equation of state for soil nitrogen', *Journal of Physical Chemistry* 34: 1053–1057.

Jenny, H. (1941) *Factors of Soil Formation: A System of Quantitative Pedology*, New York: McGraw-Hill.

Jenny, H. (1958) 'The role of the plant factor in pedogenic functions', *Ecology* 39: 5–16.

Jenny, H. (1961) 'Derivation of state factor equations of soil and ecosystems', *Soil Science Society of America Proceedings* 25: 385–388.

Jenny, H. (1965) 'Tessera and pedon', *Soil Survey Horizons* 6: 8–9.

Jenny, H. (1980) *The Soil Resource: Origin and Behavior* (Ecological Studies, vol. 37), New York: Springer.

Jenny, H. and Leonard, C. D. (1934) 'Functional relationships between soil properties and rainfall', *Soil Science* 38: 363–381.

Jenny, H., Salem, A. E., and Wallis, J. R. (1968) 'Interplay of soil organic matter and soil fertility with state factors and soil properties', in P. Salviucci (ed.) *Study Week on Organic Matter and Soil Fertility*, pp. 5–37, The Vatican: Pontificiae Academiae Scientarius, Scripta Varia 32; New York: John Wiley & Sons.

Joffe, J. S. (1949) *Pedology*, 2nd edn, with an introduction by C. F. Marbut, New Brunswick, New Jersey: Pedology Publications.

Johnson, D. L. (1980) 'Problems in the land vertebrate zoogeography of certain islands and the swimming powers of elephants', *Journal of Biogeography* 7: 383–398.

Johnson, D. L. (1989) 'Subsurface stone lines, stone zones, artifact-manuport layers and biomantles produced by bioturbation via pocket gophers (*Thomomys bottae*)', *American Antiquity* 54: 370–389.

Johnson, D. L. (1990) 'Biomantle evolution and the redistribution of earth materials and artifacts', *Soil Science* 149: 84–101.

Johnson, D. L. and Rockwell, T. K. (1982) 'Soil geomorphology: theory, concepts and principles with examples and applications on alluvial and marine terraces in coastal California', *Geological Society of America, Programs with Abstracts* 14: 176.

Johnson, D. L. and Watson-Stegner, D. (1987) 'Evolution model of pedogenesis', *Soil Science* 143: 349–366.

Johnson, D. L., Keller, E. A., and Rockwell, T. K. (1990) 'Dynamic pedogenesis: new views on some key concepts, and a model for interpreting Quaternary soils', *Quaternary Research* 33: 306–319.

Johnson, M. P. and Simberloff, D. S. (1974) 'Environmental determinants of island species numbers in the British Isles', *Journal of Biogeography* 1: 149–154.

Johnson, N. K. (1975) 'Controls of number of bird species on montane islands in the Great Basin', *Evolution* 29: 545–567.

Johnston, R. F. and Selander, R. K. (1971) 'Evolution in the house sparrow. II. Adaptive differentiation in North American populations', *Evolution* 25: 1–28.

Kachanoski, R. G. (1988) 'Processes in soils – from pedon to landscape', in T. Rosswall, R. G. Woodmansee, and P. G. Risser (eds) *Scales and Global Change: Spatial and Temporal Variability in Biospheric and Geospheric Processes* (SCOPE 35), pp. 153–177, Chichester: John Wiley & Sons.

Kadlec, R. H. and Hammer, D. E. (1988) 'Modeling nutrient behaviour in wetlands', *Ecological Modelling* 40: 37–66.

Kauffman, S. A. (1992) *The Origins of Order: Self-Organization and Selection in Evolution*, New York: Oxford University Press.

Keddy, P. A. (1991) 'Working with heterogeneity: an operator's guide to environmental gradients', in J. Kolasa and S. T. A. Pickett (eds) *Ecological Heterogeneity* (Ecological Studies, vol. 86), pp. 181–201, New York: Springer.

Kennedy, M. L., Leberg, P. L., and Baumgardner, G. D. (1986) 'Morphological variation in the coyote, *Canis latrans*, in the southern United States', *The Southwestern Naturalist* 31: 139–148.

Kessell, S. R. (1976) 'Gradient modeling: a new approach to fire modeling and wilderness resource management', *Environmental Management* 1: 39–48.

Kessell, S. R. (1979) *Gradient Modeling, Resource and Fire Management*, New York Springer.

Kikkawa, J. and Pearse, K. (1969) 'Geographical distribution of land-birds in Australia – a numerical example', *Australian Journal of Zoology* 17: 821–840.

King, G. J., Acton, D. F., and St. Arnauld, R. J. (1983) 'Soil landscape analysis in relation to soil distribution and mapping in the Weyburn Association', *Canadian Journal of Soil Science* 63: 657–670.

Kirchner, J. W. (1991) 'The Gaia hypotheses: are they testable? Are they useful?', in S. H. Schneider and P. J. Boston (eds) *Scientists on Gaia*, pp. 38–46, Cambridge, Massachusetts and London, England: the MIT Press.

Kirkby, M. J. (1989) 'A model to estimate the impact of climatic change on hillslope and regolith form', *Catena* 16: 321–341.

Kitchener, A. (1993) 'Justice at last for the dodo', *New Scientist* 139(1888): 24–27.

Knuteson, J. A., Richardson, J. L., Patterson, D. D., and Prunty, L. (1989) 'Pedogenic carbonate in a calciaquoll with a recharge wetland', *Soil Science Society of America Journal* 53: 495–499.

Koch, P. L. (1986) 'Clinal geographic variation in mammals: implications for the study of chronoclines', *Paleobiology* 12: 269–281.

Kohn, D. D. and Walsh, D. M. (1994) 'Plant species richness – the effect of island size and habitat diversity', *Journal of Ecology* 82: 367–377.

Koinov, V. Y., Boneva, K. K., and Djokova, M. Y. (1972) 'Certain geochemical features of eluvial landscapes on the most important soil-forming rocks in Bulgaria', *C. R. Acad. Agric. G. Dimitrov, Sofia* 5: 23–27.

Köppen, W. P. (1931) *Grundriss der Klimakunde. Zweite, Verbesserte Auflage der Klimate der Erde,* Berlin: Walter de Gruyter.

Kozlowski, T. T. and Ahlgren, C. E. (eds) (1974) *Fire and Ecosystems,* New York: Academic Press.

Kratter, A. W. (1992) 'Montane avian biogeography in southern California and Baja California', *Journal of Biogeography* 19: 269–283.

Krause, H. H., Rieger, S., and Wilde, S. A. (1959) 'Soils and forest growth in the Tanana Watershed of interior Alaska', *Ecology* 40: 492–495.

Kreznor, W. R., Olson, K. R., Banwart, W. L., and Johnson, D. L. (1989) 'Soil, landscape, and erosion relationships in a northwestern Illinois watershed', *Soil Science Society of America Journal* 53: 1763–1771.

Krummel, J. R., Gardner, R. H., Sugihara, G., and O'Neill, R. V. (1987) 'Landscape patterns in a disturbed environment', *Oikos* 48: 321–324.

Kump, L. R. and Garrels, R. M. (1986) 'Modeling atmospheric O_2 in the global sedimentary redox cycle', *American Journal of Science* 286: 337–360.

Lack, D. (1969) 'The numbers of bird species on islands', *Bird Study* 16: 193–209.

Lack, D. (1970) 'Island birds', *Biotropica* 2: 29–31.

LaMarche, V. C. Jr and Hirschboeck, K. K. (1984) 'Frost rings in trees as records of major volcanic eruptions', *Nature* 307: 121–126.

Lamarck, J.-B. P. A. de Monet, Chevalier de (1802) *Hydrogéologie; ou, Recherches sur l'Influence qu'ont les Eaux sur la Surface du Globe terrestre; sur les Causes de l'Existence du Bassin des Mers, de son Déplacement et de son Transport successif sur les Différens Points de la Surface de ce Globe; enfin sur les Changemens que les Corps vivans exercent sur la Nature et l'État de cette Surface,* Paris: Chez l'Auteur.

Lasaga, A. C. (1983) 'Dynamic treatment of geochemical cycles: global kinetics, in A. C. Lasaga and R. J. Kirkpatrick (eds) *Kinetics of Geochemical Processes* (Reviews in Mineralogy, vol. 8), pp. 69–110, Washington, DC: Mineralogical Society of America.

Laszlo, E. (1993) *The Creative Cosmos: A Unified Science of Matter, Life and Mind,* Edinburgh: Floris Books.

Lenihan, J. M. (1993) 'Ecological response surfaces for North American boreal tree species and their use in forest classification', *Journal of Vegetation Science* 4: 667–680.

Lenihan, J. M. and Neilson, R. P. (1993) 'A rule-based vegetation formation model for Canada', *Journal of Biogeography* 20: 615–628.

Levin, S. A. (1974) 'Dispersion and population interactions', *American Naturalist* 108: 207–228.

Levin, S. A. (1976) 'Population dynamics models in heterogeneous environments', *Annual Review of Ecology and Systematics* 7: 287–310.

Levin, S. A. (1988) 'Pattern, scale, and variability: an ecological perspective', in A. Hastings (ed.) *Community Ecology,* pp. 1–12, Berlin: Springer.

Levin, S. A. (1989) 'Challenges in the development of a theory of community structure and ecosystem function', in J. Roughgarden, R. M. May, and S. A. Levin (eds) *Perspectives in Ecological Theory,* pp. 242–255, Princeton, New Jersey: Princeton University Press.

Levin, S. A. (1991) 'The mathematics of complex systems', in H. A. Mooney, D. W. Schindler, E.-D. Schulze, and B. H. Walker (eds) *Ecosystems Experiments* (SCOPE 45), pp. 215–226, Chichester: John Wiley & Sons.

Levin, S. A., Moloney, K., Buttel, L., and Castillo-Chavez, C. (1989) 'Dynamic models of ecosystems and epidemics', *Future Generation Computer Systems* 5: 265–274.

Lewin, R. (1993) *Complexity: Life at the Edge of Chaos*, London: J. M. Dent.

Liebig, J. (1840) *Organic Chemistry and its Application to Agriculture and Physiology*, English edn edited by L. Playfair and W. Gregory, London: Taylor and Walton.

Lindsey, C. C. (1966) 'Body sizes of poikilotherm vertebrates at different latitudes', *Evolution* 20: 456–465.

Linton, D. L. (1949) 'The delimition of morphological regions', *Transactions of the Institute of British Geographers, Publication* 14: 86–87.

Lipman, P. W. and Mullineaux, D. R. (1981) 'The 1980 eruptions of Mt. St. Helens, Washington', *United States Geological Survey Professional Paper 1250*.

Litaor, M. I. (1992) 'Aluminium mobility along a geochemical catena in an alpine watershed, Front Range, Colorado', *Catena* 19: 1–16.

Lomolino, M. V. (1986) 'Mammalian community structure on islands: the importance of immigration, extinction and integrative effects', *Biological Journal of the Linnean Society* 28: 1–21.

Lomolino, M. V., Brown, J. H., and Davis, R. (1989) 'Island biogeography of montane forest mammals in the American Southwest', *Ecology* 70: 180–194.

Lord, R. D. (1960) 'Litter size and latitude in North American mammals', *American Midland Naturalist* 64: 488–499.

Lotka, A. J. (1925) *Elements of Physical Biology*, Baltimore: Williams & Wilkins. Reprinted with corrections and bibliography as *Elements of Mathematical Biology*, New York: Dover, 1956.

Lotspeich, F. B. and Smith, H. W. (1953) 'Soils of the Palouse loess: I. The Palouse catena', *Soil Science* 76: 467–480.

Lundelius, E. L. Jr, Graham, R. W., Anderson, E., Guilday, J., Holman, J. A., Steadman, D., and Webb, S. D. (1983) 'Terrestrial vertebrate faunas', in S. C. Porter (ed.) *Late-Quaternary Environments of the United States, Volume 1, The Late Pleistocene*, pp. 311–353, London: Longman.

Lutz, H. J. and Chandler, R. F. (1946) *Forest Soils*, New York: John Wiley & Sons.

MacArthur, R. H. and Wilson, E. O. (1963) 'An equilibrium theory of insular zoogeography', *Evolution* 17: 373–387.

MacArthur, R. H. and Wilson, E. O. (1967) *The Theory of Island Biogeography*, Princeton, New Jersey: Princeton University Press.

McCune, B. and Allen, T. F. H. (1985a) 'Will similar forests develop on similar sites?', *Canadian Journal of Botany* 63: 367–376.

McCune, B. and Allen, T. F. H. (1985b) 'Forest dynamics in Bitterroot Canyons, Montana', *Canadian Journal of Botany* 63: 377–383.

MacFadden, B. J. (1980) 'Rafting animals or drifting islands? Antillean insectivores *Nesophontes* and *Solenodon*', *Journal of Biogeography* 7: 11–22.

McIntosh, R. P. (1991) 'Concept and terminology of homogeneity and heterogeneity in ecology', in J. Kolasa and S. T. A. Pickett (eds) *Ecological Heterogeneity* (Ecological Studies, vol. 86), New York: Springer.

McLaren, D. J. (1988a) 'Detection and significance of mass killings', in N. J. McMillan, A. F. Embry, and D. J. Glass (eds) *Devonian of the World, Volume III: Paleontology, Paleoecology, and Biostratigraphy*, pp. 1–7, Calgary, Alberta: Canadian Society of Petroleum Geologists.

McLaren, D. J. (1988b) 'Rare events in geology', *Eos* 69: 24–25.

McLaren, D. J. (1970) 'Presidential address: Time, life, and boundaries', *Journal of Paleontology* 44: 801–815.

McLaren, D. J. and Goodfellow, W. D. (1990) 'Geological and biological consequences of giant impacts', *Annual Review of Earth and Planetary Sciences* 18: 123–171.

McLean, D. M. (1981) 'A test of terminal Mesozoic "catastrophe" ', *Earth and Planetary Science Letters* 53: 103–108.

McNaughton, S. J. (1993) 'Biodiversity and function of grazing ecosystems', in E.-D. Schulze and H. A. Mooney (eds) *Biodiversity and Ecosystem Function* (Ecological Studies, vol. 99), pp. 361–383, New York: Springer.

Major, J. (1951) 'A functional factorial approach to plant ecology', *Ecology* 32: 392–412.

Malanson, G. P., Butler, D., and Walsh, S. J. (1990) 'Chaos theory in physical geography', *Physical Geography* 11: 293–304.

Malanson, G. P., Butler, D., and Georgakakos, K. P. (1992) 'Nonequilibrium geomorphic processes and deterministic chaos', *Geomorphology* 5: 311–322.

Malhotra, A. and Thorpe, R. S. (1991a) 'Microgeographic variation in *Anolis oculatus*, on the island of Dominica, West Indies', *Journal of Evolutionary Biology* 4: 321–335.

Malhotra, A. and Thorpe, R. S. (1991b) 'Experimental detection of rapid evolutionary response in natural lizard populations', *Nature* 353: 347–348.

Marbut, C. F. (1927) 'A scheme for soil classification', *Proceedings and Papers of the First International Congress of Soil Science* 4: 1–31.

Marianatos, S. (1939) 'The volcanic destruction of Minoan Crete', *Antiquity* 13: 425–439.

Marion, G. M. (1989) 'Correlation between long-term pedogenic $CaCO_3$ formation rate and modern precipitation in deserts of the American Southwest', *Quaternary Research* 32: 291–295.

Marshall, J. K. (1978) 'Factors limiting the survival of *Corynephorus canescens* (L.) Beauv. in Great Britain at the northern edge of its distribution', *Oikos* 19: 206–216.

Matthews, J. A. (1979a) 'The vegetation of the Storbreen gletschervorfeld, Jotunheimen, Norway. I. Introduction and approaches involving classification', *Journal of Biogeography* 6: 17–47.

Matthews, J. A. (1979b) 'The vegetation of the Storbreen gletschervorfeld, Jotunheimen, Norway. II. Approaches involving ordination and general conclusions', *Journal of Biogeography* 6: 133–167.

Matthews, J. A. (1979c) 'A study of the variability of some successional and climax assemblage types using multiple discriminant analysis', *Journal of Ecology* 67: 225–271.

Matthews, J. A. (1992) *The Ecology of Recently-Deglaciated Terrain: A Geoecological Approach to Glacier Forelands and Primary Succession*, Cambridge: Cambridge University Press.

Mattson, S. (1938) 'The constitution of the pedosphere', *Annals of the Agricultural College of Sweden* 5: 261–276.

May, R. M. (1973) *Stability and Complexity in Model Ecosystems*, Princeton, New Jersey: Princeton University Press.

May, R. M. (1976) 'Simple mathematical models with very complicated dynamics', *Nature* 261: 459–467.

Mayr, E. (1942) *Systematics and the Origin of Species*, New York: Columbia University Press.

Mayr, E. (1956) 'Geographical character gradients and climatic adaptation', *Evolution* 20: 105–108.

Mayr, E. (1965) 'What is a fauna?', *Zoologische Jahrbücher (Systematik, Ökologie und Geographie der Tiere)* 92: 437–486.

Meentemeyer, V. (1978) 'Macroclimate and lignin control of litter decomposition rates', *Ecology* 59: 465–472.

Meentemeyer, V. (1984) 'The geography of organic decomposition rates', *Annals of the Association of American Geographers* 74: 551–560.

Meentemeyer, V. (1989) 'Geographical perspectives of space, time, and scale', *Landscape Ecology* 3: 163–173.

Meentemeyer, V. and Box, E. O. (1987) 'Scale effects in landscape studies', in M. G. Turner (ed.) *Landscape Heterogeneity and Disturbance*, pp. 15–36, New York: Springer.

Meher-Homji, V. M. (1963) *Les Bioclimats du Sub-Continent Indien et leurs Types analogues dans le Monde* (Documents pour les Cartes des Productions Végétales; Série: Généralités, vol. i, tome iv), Toulouse: Faculté des Sciences.

Melosh, H. J., Scheider, N. M., Zahnle, K. J., and Lathan, D. (1990) 'Ignition of global wildfires at the Cretaceous/Tertiary boundary', *Nature* 343: 251–254.

Merriam, C. H. (1894) 'Laws of temperature control of the geographic distribution of terrestrial animals and plants', *National Geographic Magazine* 6: 229–238.

Meyer, G. A., Wells, S. G., Balling, R. C., Jr, and Jull, A. J. T. (1992) 'Response of alluvial systems to fire and climate change in Yellowstone National Park', *Nature* 357: 147–150.

Milchunas, D. G. and Lauenroth, W. K. (1993) 'Quantitative effects of grazing on vegetation and soils over a global range of environments', *Ecological Monographs* 63: 327–366.

Mill, H. R. (1899) 'Geography: principles and progress', in H. R. Mill (ed.) *The International Geography*, pp. 2–13, London: George Newnes.

Milne, G. (1935a) 'Some suggested units of classification and mapping, particularly for East African soils', *Soil Research* 4: 183–198.

Milne, G. (1935b) 'Composite units for the mapping of complex soil associations', *Transactions of the Third International Congress of Soil Science, Oxford, England, 1935* 1: 345–347.

Milne, G. (1936) 'Normal erosion as a factor in soil profile development', *Nature* 138: 548–549.

Moore, I. D., Grayson, R. B., and Ladson, A. R. (1991) 'Digital terrain modelling: a review of hydrological, geomorphological and biological applications', *Hydrological Processes* 5: 3–30.

Moore, I. D., Gessler, P. E., Nielsen, G. A., and Peterson, G. A. (1993) 'Soil attribute prediction using terrain analysis', *Soil Science Society of America Journal* 57: 443–452.

Morison, C. G. T. (1949) 'The catena concept and the classification of tropical soils', in *Proceedings of the First Commonwealth Conference on Tropical and Sub-Tropical Soils, 1948* (Commonwealth Bureau of Soil Science, Technical Communication No. 46), Harpenden, England: Commonwealth Bureau of Soil Science.

Mueller-Dombois, D. and Bridges, K. W. (1981) 'Introduction', in D. Mueller-Dombois, K. W. Bridges, and H. L. Carson (eds) *Island Ecosystems: Biological Organization in Selected Hawaiian Communities* (US/IBP Synthesis Series, vol. 15), pp. 35–76, Stroudsberg, Pennsylvania and Woods Hole, Massachusetts: Hutchinson Ross.

Mueller-Dombois, D., Spatz, G., Conant, S., Tomich, P. Q., Radovsky, F. J., Tenorio, J. M., Gagné, W., Brennan, B. M., Mitchell, W. C., Springer, D., Samuelson, G. A., Gressit, J. L., Steffan, W. A., Paik, Y. K., Sung, K. C., Hardy, D. E., Delfinado, M. D., Fujii, D., Doty, M. S., Watson, L. J., Stoner, M. F., and Baker, G. E. (1981)

'Altitudinal distribution of organisms along an island mountain transect', in D. Mueller-Dombois, K. W. Bridges, and H. L. Carson (eds) *Island Ecosystems: Biological Organization in Selected Hawaiian Communities* (US/IBP Synthesis Series, vol. 15), pp. 77–180, Stroudsberg, Pennsylvania and Woods Hole, Massachusetts: Hutchinson Ross.

Muhs, D. R. (1982) 'The influence of topography on the spatial variability of soils in Mediterranean climates', in C. E. Thorn (ed.) *Space and Time in Geomorphology*, pp. 269–284, London: George Allen & Unwin.

Myklestad, Å. (1993) 'The distribution of *Salix* species in Fennoscandia – a numerical analysis', *Ecography* 16: 329–344.

Myklestad, Å. and Birks, H. J. B. (1993) 'A numerical analysis of the distribution patterns of *Salix* L. species in Europe', *Journal of Biogeography* 20: 1–32.

Napier, W. M. and Clube, S. V. M. (1979) 'A theory of terrestrial catastrophism', *Nature* 282: 455–459.

Neilson, R. P. and Wullstein, L. H. (1983) 'Biogeography of two southwest American oaks in relation to atmospheric dynamics', *Journal of Biogeography* 10: 275–297.

Neilson, R. P. and Wullstein, L. H. (1986) 'Microhabitat affinities of Gambel oak seedlings', *Great Basin Naturalist* 46: 294–298.

Neilson, R. P., King, G. A., DeVelice, R. L., and Lenihan, J. M. (1992) 'Regional and local vegetation patterns: the responses of vegetation diversity to subcontinental air masses', in A. J. Hansen and F. di Castri (eds) *Landscape Boundaries: Consequences for Biotic Diversity and Ecological Flows* (Ecological Studies, vol. 92), pp. 129–149, New York: Springer.

Nevo, E. (1986) 'Mechanisms of adaptive speciation at the molecular and organismal levels', in S. Karlin and E. Nevo (eds) *Evolutionary Processes and Theory*, pp. 439–474, New York: Academic Press.

Nikiforoff, C. C. (1935) 'Weathering and soil formation', *Transactions of the Third International Congress of Soil Science, Oxford, England, 1935* 1: 324–326.

Nininger, H. H. (1942) 'Cataclysm and evolution', *Popular Astronomy* 50: 270–272.

Nix, H. (1982) 'Environmental determinants of biogeography and evolution in Terra Australis', in W. R. Barker and P. J. M. Greenslade (eds) *Evolution of the Flora and Fauna of Arid Australia*, pp. 47–66, Adelaide: Peacock Publications.

Nunn, P. D. (1993) *Oceanic Islands*, Oxford: Blackwell.

Officer, C. B. (1993) 'Victims of volcanoes', *New Scientist* 137(1861): 34–38.

Officer, C. B., Hallam, A., Drake, C. L., and Devine, J. D. (1987) 'Late Cretaceous and paroxysmal Cretaceous/Tertiary extinctions', *Nature* 326: 143–149.

Ogden, J. and Powell, J. A. (1979) 'A quantitative description of the forest vegetation in an altitudinal gradient in the Mount Field National Park, Tasmania, and a discussion of its history and dynamics', *Australian Journal of Ecology* 4: 293–325.

Olson, S. L. (1975) 'Paleornithology of St. Helena Island, South Atlantic Ocean', *Smithsonian Contributions to Paleobiology* 23: 1–43.

Onda, Y. (1992) 'Influence of water storage capacity in the regolith zone on hydrological characteristics, slope processes, and slope form', *Zeitschrift für Geomorphologie* NF 36: 165–178.

O'Neill, R. V., Gardner, R. H., Milne, B. T., Turner, M. G., and Jackson, B. (1991) 'Heterogeneity and spatial hierarchies', in J. Kolasa and S. T. A. Pickett (eds) *Ecological Heterogeneity* (Ecological Studies, vol. 86), pp. 85–96, New York: Springer.

Öpik, E. J. (1958) 'On the catastrophic effects of collisions with celestial bodies', *Irish Astronomical Journal* 5: 34–36.

Osburn, R. C., Dublin, L. I., Shimer, H. W., and Lull, R. S. (1903) 'Adaptation to aquatic, arboreal, fossorial, and cursorial habits in mammals', *American Naturalist*

37: 651–665, 731–736, 819–825; 38: 322–332.

Ovalles, F. A. and Collins, M. E. (1986) 'Soil–landscape relationships and soil variability in North Central Florida', *Soil Science Society of America Journal* 50: 401–408.

Overpeck, J. T., Rind, D., and Goldberg, R. (1990) 'Climate-induced changes in forest disturbance and vegetation', *Nature* 343: 51–53.

Owen, D. F. (1980) 'How plants may benefit from the animals that eat them', *Oikos* 35: 230–235.

Pang, K. D. and Chou, H.-H. (1985) 'Three very large volcanic eruptions in antiquity and their effects on the climate of the ancient world', *Eos* 67: 880–881 (Abstract).

Parton, W. J., Schimel, D. S., Cole, C. V., and Ojima, D. S. (1987) 'Analysis of factors controlling soil organic matter levels in Great Plains grasslands', *Soil Science Society of America Journal* 51: 1173–1179.

Parton, W. J., Cole, C. V., Stewart, J. W. B., Ojima, D. S., and Schimel, D. S. (1988) 'Simulating regional patterns of soil C, N, and P dynamics in the U.S. central grasslands region', *Biogeochemistry* 5: 109–131.

Patterson, B. D. (1984) 'Mammalian extinction and biogeography in the Southern Rocky Mountains', in M. H. Nitecki (ed.) *Extinctions*, pp. 247–293, Chicago: University of Chicago Press.

Patterson, B. D. (1987) 'The principle of nested subsets and its implications for biological conservation', *Conservation Biology* 1: 323–334.

Patterson, B. D. (1991) 'The integral role of biogeographic theory in the conservation of tropical rain forest diversity', in M. A. Mares and D. J. Schmidly (eds) *Latin American Mammalogy: History, Biodiversity, and Conservation*, pp. 125–149, Norman, Oklahoma: University of Oklahoma Press.

Patterson, B. D. and Atmar, W. (1986) 'Nested subsets and the structure of insular mammalian faunas and archipelagos', *Biological Journal of the Linnean Society* 28: 65–82.

Patterson, B. D. and Brown, J. H. (1991) 'Regionally nested patterns of species composition in granivorous rodent assemblages', *Journal of Biogeography* 18: 395–402.

Pears, N. (1985) *Basic Biogeography*, 2nd edn, Harlow: Longman.

Pedersen, K. (1993) 'The deep subterranean biosphere', *Earth-Science Reviews* 34: 243–260.

Pennock, D. J. and de Jong, E. (1987) 'The influence of slope curvature on soil erosion and deposition in hummock terrain', *Soil Science* 144: 209–217.

Pérez, F. L. (1987) 'Soil moisture and the upper altitudinal limit of giant paramo rosettes', *Journal of Biogeography* 14: 173–186.

Pérez, F. L. (1989) 'Some effects of giant Andean stem-rosettes on ground micro-climate, and their ecological significance', *International Journal of Biometeorology* 33: 131–135.

Pérez, F. L. (1991) 'Soil moisture and the distribution of giant Andean rosettes on talus slopes of a desert paramo', *Climate Research* 1: 217–231.

Perring, F. H. (1958) 'A theoretical approach to a study of chalk grassland', *Journal of Ecology* 46: 665–679.

Perring, F. H. and Walters, S. M. (1962) *Atlas of the British Flora*, London: Nelson.

Phillips, J. D. (1990) 'Relative ages of wetland and upland surfaces as indicated by pedogenic development', *Physical Geography* 11: 363–378.

Phillips, J. D. (1992) 'Qualitative chaos in geomorphic systems, with an example from wetland response to sea level rise', *Journal of Geology* 100: 365–374.

Phillips, J. D. (1993a) 'Instability and chaos in hillslope evolution', *American Journal of Science* 293: 25–48.

Phillips, J. D. (1993b) 'Spatial-domain chaos in landscapes', *Geographical Analysis* 25: 101–117.

Phillips, J. D. (1993c) 'Stability implications of the state factor model of soils as a nonlinear dynamical system', *Geoderma* 58: 1–15.

Pickett, S. T. A. and White, P. S. (eds) (1985) *The Ecology of Natural Disturbance and Patch Dynamics*, Orlando, Florida: Academic Press.

Pielou, E. C. (1977) 'The latitudinal spans of seaweed species and their patterns of overlap', *Journal of Biogeography* 4: 299–311.

Pigott, C. D. (1974) 'The response of plants to climate and climatic change', in F. Perring (ed.) *The Flora of a Changing Britain*, pp. 32–44, London: Classey.

Pigott, C. D. and Huntley, J. P. (1981) 'Factors controlling the distribution of *Tilia cordata* at the northern limits of its geographical range. III. Nature and causes of seed sterility', *New Phytologist* 87: 817–839.

Pimm, S. L. (1982) *Food Webs*, London and New York: Chapman and Hall.

Pimm, S. L. (1992) *Balance of Nature? Ecological Issues in the Conservation of Species and Communities*, Chicago, Illinois: Chicago University Press.

Polunin, N. and Grinevald, J. (1988) 'Vernadsky and biospheral ecology', *Environmental Conservation* 15: 117–122.

Polynov, B. B. (1935) 'Types of weathering crust', *Transactions of the Third International Congress of Soil Science, Oxford, England, 1935* 1: 327–330.

Polynov, B. B. (1937) *The Cycle of Weathering*, translated from the Russian by A. Muir, with a foreword by W. G. Ogg, London: Thomas Murby.

Prentice, I. C., Cramer, W., Harrison, S. P., Leemans, R., Monserud, R. A., and Solomon, A. M. (1992) 'A global biome model based on plant physiology and dominance, soil properties and climate', *Journal of Biogeography* 19: 117–134.

Preston, F. W. (1962) 'The canonical distribution of commonness and rarity', *Ecology* 43: 185–215, 410–432.

Prinn, R. G. and Fegley, B. Jr (1987) 'Bolide impacts, acid rain, and biospheric traumas at the Cretaceous–Tertiary boundary', *Earth and Planetary Science Letters* 83: 1–15.

Puccia, C. J. and Levins, R. (1985) *Qualitative Modeling of Complex Systems*, Cambridge, Massachusetts: Harvard University Press.

Puccia, C. J. and Levins, R. (1991) 'Qualitative modeling in ecology: loop analysis, signed digraphs, and time averaging', in P. A. Fishwick and P. A. Luker (eds) *Qualitative Simulation Modeling and Analysis* (Advances in Simulation, vol. 5), pp. 119–143, New York: Springer.

Rampino, M. R., Stothers, R. B., and Self, S. (1985) 'Climatic effects of volcanic eruptions', *Nature* 313: 272.

Rampino, M. R., Self, S., and Stothers, R. B. (1988) 'Volcanic winters', *Annual Review of Earth and Planetary Sciences* 16: 73–99.

Rastall, R. H. (1941) *Lake and Rastall's Textbook of Geology*, revised by R. H. Rastall, 5th edn, London: Edward Arnold.

Raunkiaer, C. (1934) *The Life Forms of Plants and Statistical Plant Geography*, Oxford: Oxford University Press.

Raup, D. M. (1990) 'Impact as a general cause of extinction: a feasibility test', in V. L. Sharpton and P. D. Ward (eds) *Global Catastrophes in Earth History; An Interdisciplinary Conference on Impacts, Volcanism, and Mass Mortality*, pp. 27–32, Geological Society of America Special Paper 247.

Raup, D. M. and Jablonski, D. (1993) 'Geography of end-Cretaceous marine bivalve extinctions', *Science* 260: 971–973.

Raup, H. M. (1981) 'Physical disturbance in the life of plants', in M. H. Nitecki (ed.) *Biotic Crises in Ecological and Evolutionary Time*, pp. 39–52, New York: Academic Press.

Rawlinson, P. A., Widjoya, A. H. T., Hutchinson, M. N., and Brown, G. W. (1990) 'The terrestrial vertebrate fauna of the Krakatau Islands, Sunda Strait, 1883–1986', *Philosophical Transactions of the Royal Society of London* 328B: 3–28.

Ray, C. (1960) 'The application of Bergmann's and Allen's rules to poikilotherms', *Journal of Morphology* 106: 85–108.

Read, J. and Hill, R. S. (1988) 'Comparative responses to temperature of the major canopy species of Tasmanian cool temperate rainforest and their ecological significance. I. Foliar frost resistance', *Australian Journal of Botany* 36: 131–143.

Read, J. and Hill, R. S. (1989) 'The response of some Australian temperate rain forest tree species to freezing temperatures and its biological significance', *Journal of Biogeography* 16: 21–27.

Read, J. and Hope, G. S. (1989) 'Foliar resistance of some evergreen tropical and extratropical Australasian *Nothofagus* species', *Australian Journal of Botany* 37: 361–373.

Rebertus, R. A., Doolittle, J. A., and Hall, R. L. (1989) 'Landform and stratigraphic influences on variability of loess thickness in northern Delaware', *Soil Science Society of America Journal*, 53: 843–847.

Remmert, H. (1991) 'The mosaic-cycle concept of ecosystems – an overview', in H. Rennert (ed.) *The Mosaic-Cycle Concept in Ecosystems* (Ecological Studies, vol. 85), pp. 1–21, New York: Springer.

Renne, P. R. and Basu, A. R. (1991) 'Rapid eruption of the Siberian traps flood basalts at the Permo-Triassic boundary', *Science* 253: 176–179.

Rensch, B. (1932) 'Über die Abhängigkeit der Grösse, des relativen Gewichtes und der Oberflächenstruktur der Lanschnecknschalen von dem Umweltsfaktoren', *Zeitschrift für Morphologie und Ökologie* 25: 757–807.

Rhodes, R. S. II (1984) 'Paleoecological and regional paleoclimatic implications of the Farmdalian Craigmile and Woodfordian Waubonsie mammalian local faunas, southwestern Iowa, *Illinois State Museum Report of Investigations* 40: 1–51.

Richardson, D. M. and van Wilgen, B. W. (1992) 'Ecosystem, community and species response to fire in mountain fynbos: conclusions from the Swartboskloof experiment', in B. W. van Wilgen, D. M. Richardson, F. J. Kruger, and H. J. van Hensbergen (eds) *Fire in South African Mountain Fynbos: Ecosystem, Community and Species Response at Swartboskloof* (Ecological Studies, vol. 93), pp. 273–284, New York: Springer.

Richardson, J. L. and Edmonds, W. J. (1987) 'Linear regression estimations of Jenny's relative effectiveness of state factor equations', *Soil Science* 144: 203–208.

Richardson, J. L., Wilding, L. P., and Daniels, R. B. (1992) 'Recharge and discharge of groundwater in aquic conditions illustrated with flownet analysis', *Geoderma* 53: 65–78.

Ricklefs, R. E. and Cox, G. W. (1972) 'Taxon cycles in the West Indian avifauna', *American Naturalist* 106: 195–219.

Ricklefs, R. E. and Cox, G. W. (1978) 'Stage of taxon cycle, habitat distribution, and population density in the avifauna of the West Indies', *American Naturalist* 112: 875–895.

Robinson, G. W. (1936) 'Normal erosion as a factor in soil profile development', *Nature* 137: 950.

Romme, W. H. (1982) 'Fire and landscape diversity in subalpine forests of Yellowstone National Park', *Ecological Monographs* 52: 199–221.

Romme, W. H. and Despain, D. G. (1989) 'The Yellowstone fires', *Scientific American* 261: 37–46.

Rorison, I. H., Sutton, F., and Hunt, R. (1986) 'Local climate, topography and plant growth in Lathkill Dale NNR. I. A twelve-year summary of solar radiation and temperature', *Plant, Cell, and Environment* 9: 49–56.

Rose, A. W., Hawkes, H. E., and Webb, J. S. (1979) *Geochemistry in Mineral Exploration*, 2nd edn, London: Academic Press.

Rose, W. I. and Chesner, C. A. (1990) 'Worldwide dispersal of ash and gases from earth's largest known eruption: Toba, Sumatra, 75 ka', *Palaeogeography, Palaeoclimatology, Palaeoecology (Global and Planetary Change Section)* 89: 269–275.

Rosenzwieg, M. L. (1966) 'Community structure in sympatric carnivora', *Journal of Mammalogy* 47: 602–612.

Rosenzweig, M. L. (1968) 'The strategy of body size in mammalian carnivores', *The American Midland Naturalist* 80: 299–315.

Roy, A. G., Jarvis, R. S., and Arnett, R. R. (1980) 'Soil–slope relationships within a drainage basin', *Annals of the Association of American Geographers* 70: 397–412.

Ruhe, R. V. (1983) 'Aspects of Holocene pedology in the United States', in H. E. Wright Jr. (ed.) *Late-Quaternary Environments of the United States, Volume 2, The Holocene*, pp. 12–25, London: Longman.

Ruhe, R. V. (1984) 'Loess-derived soils, Mississippi Valley region: II. Soil–climate system', *Soil Science Society of America Journal* 48: 864–867.

Ruhe, R. V. and Walker, P. H. (1968) 'Hillslopes and soil formation. I. Open systems', *Transactions of the Ninth International Congress of Soil Science, Adelaide* 4: 551–560.

Russell, D. A. (1979) 'The enigma of the extinction of the dinosaurs', *Annual Review of Earth and Planetary Sciences* 7: 163–182.

Rykiel, E. J., Jr, Coulson, R. N., Sharpe, P. J. H., Allen, T. F. H., and Flamm, R. O. (1988) 'Disturbance propagation by bark beetles as an episodic landscape phenomenon', *Landscape Ecology* 1: 129–139.

Salisbury, E. J. (1926) 'The geographical distribution of plants in relation to climatic factors', *Geographical Journal* 57: 312–335.

Samways, M. J. (1989) 'Climate diagrams and biological control: an example from the areography of the ladybird *Chilocorus nigritus* (Fabricius, 1798) (Insecta, Coleoptera, Coccinellidae)', *Journal of Biogeography* 16: 345–351.

Sauer, J. D. (1969) 'Oceanic islands and biogeographic theory: a review', *Geographical Review* 59: 582–593.

Schaetzl, R. J., Burns, S. F., Johnson, D. L., and Small, T. W. (1989a) 'Tree uprooting: a review of impacts on forest ecology', *Vegetatio* 79: 165–176.

Schaetzl, R. J., Johnson, D. L., Burns, S. F., and Small, T. W. (1989b) 'Tree uprooting: review of terminology, process, and environmental implications', *Canadian Journal of Forest Research* 19: 1–11.

Scheidegger, A. E. (1986) 'The catena principle in geomorphology', *Zeitschrift für Geomorphologie* 30: 257–273.

Schlesinger, W. H., DeLucia, E. H., and Billings, W. D. (1989) 'Nutrient-use efficiency of woody plants on contrasting soils in the western Great Basin, Nevada', *Ecology* 70: 105–113.

Schneider, S. H. and Boston, P. J. (eds) (1991) *Scientists on Gaia*, Cambridge, Massachusetts and London, England: the MIT Press.

Selles, F., Karamanos, R. E., and Kachanoski, R. G. (1986) 'The spatial variability of nitrogen-15 and its relation to the variability of other soil properties', *Soil Science Society of America Journal* 50: 105–110.

Shachak, M. and Brand, S. (1991) 'Relations among spatiotemporal heterogeneity, population abundance, and variability in a desert', in J. Kolasa and S. T. A. Pickett (eds) *Ecological Heterogeneity* (Ecological Studies, vol. 86) pp. 202–223, New York: Springer.

Shafer, C. L. (1990) *Nature Reserves: Island Theory and Conservation Practice*,

Washington and London: Smithsonian Institution Press.

Sharpton, V. L. and Grieve, R. A. F. (1990) 'Meteorite impact, cryptoexplosion, and shock metamorphism; a perspective on the evidence at the K/T boundary', in V. L. Sharpton and P. D. Ward (eds) *Global Catastrophes in Earth History; An Interdisciplinary Conference on Impacts, Volcanism, and Mass Mortality*, pp. 301–318, Geological Society of America Special Paper 247.

Sharpton, V. L. and Ward, P. D. (eds) (1990) *Global Catastrophes in Earth History; An Interdisciplinary Conference on Impacts, Volcanism, and Mass Mortality*, Geological Society of America Special Paper 247.

Shaw, C. F. (1930) 'Potent factors in soil formation', *Ecology* 11: 239–245.

Shelford, V. E. (1911) 'Physiological animal geography', *Journal of Morphology* 22: 551–618.

Shimwell, D. W. (1971) *Description and Classification of Vegetation*, London: Sidgwick & Jackson.

Shoji S., Nanzyo, M., and Dahlgren, R. A. (1993) *Volcanic Ash Soils: Genesis, Properties and Utilization* (Developments in Science, vol. 21), Amsterdam: Elsevier.

Shugart, H. H. and Noble, J. R. (1981) 'A computer model of succession and fire response of the high altitude *Eucalyptus* forest of the Brindabella Range, Australian Capital Territory', *Australian Journal of Ecology* 6: 149–164.

Simberloff, D. S. (1976) 'Experimental zoogeography of islands: the effects of island size', *Ecology* 57: 629–648.

Simpson, M. R. and Boutin, S. (1993) 'Muskrat life history: a comparison of a northern and southern population', *Ecography* 16: 5–10.

Sklar, F. H. and Costanza, R. (1991) 'The development of dynamic spatial models for landscape ecology: a review and prognosis', in M. G. Turner and R. H. Gardner (eds) *Quantitative Methods in Landscape Ecology: the Analysis and Interpretation of Landscape Heterogeneity* (Ecological Studies, vol. 82), pp. 239–288.

Slayter, R. O. and Noble, I. R. (1992) 'Dynamics of montane treelines', in A. J. Hansen and F. di Castri (eds) *Landscape Boundaries: Consequences for Biotic Diversity and Ecological Flows* (Ecological Studies, vol. 92), pp. 346–359, New York: Springer.

Smeck, N. E. and Runge, E. C. A. (1971) 'Phosphorus availability and redistribution in relation to profile development in an Illinois landscape segment', *Soil Science Society of America Proceedings* 35: 952–959.

Smith, A. T. (1974) 'The distribution and dispersal of pikas: consequences of insular population structure', *Ecology* 55: 1112–1119.

Smith, A. T. (1980) 'Temporal changes on insular populations of the pika (*Ochotona princeps*)', *Ecology* 61: 8–13.

Smith, C. H. (1983a) 'A system of world mammal faunal regions. I. Logical and statistical derivation of the regions', *Journal of Biogeography* 10: 467–482.

Smith, C. H. (1983b) 'A system of world mammal faunal regions. II. The distance decay effect upon inter-regional affinities', *Journal of Biogeography* 10: 467–482.

Smith, W. K. and Knapp, A. K. (1990) 'Ecophysiology of high elevation forests', in C. B. Osmond, L. F. Pitelka, and G. M. Hidy (eds) *Plant Biology of the Basin and Range* (Ecological Studies, vol. 80), pp. 87–142, Berlin: Springer.

Soil Survey Staff (1975) *Soil Taxonomy: A Basic System of Soil Classification for Making and Interpreting Soil Surveys*, US Department of Agriculture, Agricultural Handbook 436, Washington, DC: US Government Printing Office.

Sondaar, P. Y. (1976) 'Insularity and its effect on mammal evolution', in M. K. Hecht, P. C. Goody, and B. M. Hecht (eds) *Major Patterns in Vertebrate Evolution*, New York: Plenum Press.

Speight, J. G. (1974) 'A parametric approach to landform regions', in E. H. Brown and R. S. Waters (eds) *Progress in Geomorphology: Papers in Honour of David L. Linton* (Institute of British Geographers, Special Publication No. 7), pp. 213–230, London: Institute of British Geographers.

Stanley, D. J. and Sheng, H. (1986) 'Volcanic shards from Santorini (Upper Minoan ash) in the Nile Delta, Egypt', *Nature* 320: 733–735.

Stephens, C. G. (1947) 'Functional synthesis in pedogenesis', *Transactions of the Royal Society of South Australia* 71: 168–181.

Stephens, E. P. (1956) 'The uprooting of trees, a forest process', *Soil Science Society of America Proceedings* 20: 113–116.

Stephenson, N. L. (1990) 'Climate control of vegetation distribution: the role of the water balance', *American Naturalist* 135: 649–670.

Stolt, M. H., Baker, J. C., and Simpson, T. W. (1993a) 'Soil–landscape relationships in Virginia: I. Soil variability and parent material uniformity', *Soil Science Society of America Journal* 57: 414–421.

Stolt, M. H., Baker, J. C., and Simpson, T. W. (1993b) 'Soil–landscape relationships in Virginia: II. Reconstruction analysis and soil genesis', *Soil Science Society of America Journal* 57: 422–428.

Stothers, R. B. (1984) 'The great Tambora eruption in 1815 and its aftermath', *Science* 224: 223–224.

Stothers, R. B. and Rampino, M. R. (1983) 'Volcanic eruptions in the Mediterranean before A.D. 630 from written and archaeological sources', *Journal of Geophysical Research* 88B: 6357–6371.

Stothers, R. B., Wolff, J. A., Self, S., and Rampino, M. R. (1986) 'Basaltic fissure eruptions, plume heights, and atmospheric aerosols', *Geophysical Research Letters* 13: 725–728.

Stoutjesdijk, P. and Barkman, J. J. (1992) *Microclimate, Vegetation and Fauna*, Uppsala, Sweden: Opulus Press.

Strughold, H. (1953) *The Green and Red Planet: A Physiological Study of the Possibility of Life on Mars*, Albuquerque, New Mexico: University of New Mexico Press.

Suess, E. (1875) *Die Entstehung der Alpen*, Wien: W. Braunmüller.

Suess, E. (1883–1909) *Das Antlitz der Erde*, 5 vols, Wien: G. Freytag.

Suess, E. (1909) *The Face of the Earth (Das Antlitz der Erde)*, vol. 4, translated by Hertha B. C. Sollas under the direction of W. J. Sollas, Oxford: Clarendon Press.

Sukachev, V. N. and Dylis, N. (1968) *Fundamentals of Forest Biogeocoenecology*, translated from the Russian by J. M. Maclennan, Edinburgh and London: Oliver & Boyd.

Sutherland, R. A., van Kessel, C., Farrell, R. E., and Pennock, D. J. (1993) 'Landscape-scale variations in soil nitrogen-15 natural abundance', *Soil Science Society of America Journal* 57: 169–178.

Swanson, D. K. (1985) 'Soil catenas on Pinedale and Bull Lake moraines, Willow Lake, Wind River Mountains, Wyoming', *Catena* 12: 329–342.

Swetnam, T. W. and Betancourt, J. L. (1990) 'Fire–Southern Oscillation relations in the Southwestern United States', *Science* 249: 1017–1020.

Tansley, A. G. (1935) 'The use and abuse of vegetational concepts and terms', *Ecology* 16: 284–307.

Tansley, A. G. (1939) *The British Isles and Their Vegetation*, Cambridge: Cambridge University Press.

Teeuw, R. M. (1989) 'Variations in the composition of gravel layers across the landscape. Examples from Sierra Leone', *Geo-Eco-Trop* 11: 151–169.

Teeuw, R. M. (1991) 'A catenary approach to the study of gravel layers and tropical landscape morphodynamics', *Catena* 18: 71–89.

Teilhard de Chardin, P. (1957) *La Vision du Passé*, Paris: Éditions du Seuil.

Teilhard de Chardin, P. (1959) *The Phenomenon of Man*, with an introduction by Sir Julian Huxley, London: Collins.

Teilhard de Chardin, P. (1969) *The Future of Man*, translated from the French by N. Denny. London: Collins.

Terborgh, J. (1971) 'Distribution on environmental gradients: theory and a preliminary interpretation of distributional patterns in the avifauna of the Cordillera Vilcabamba, Peru', *Ecology* 52: 23–40.

Terborgh, J. and Weske, J. S. (1975) 'The role of competition in the distribution of Andean birds', *Ecology* 56: 562–576.

Tewari, J. C., Rikhari, H. C., and Singh, S. P. (1989) 'Compositional and structural features of certain tree stands along an elevational gradient in central Himalaya', *Vegetatio* 85: 107–120.

Thomas, D. S. G. (1988) 'The biogeomorphology of arid and semi-arid environments', in H. A. Viles (ed.) *Biogeomorphology*, pp. 193–221, Oxford: Basil Blackwell.

Thomas, M. F., Thorp, M. B., and Teeuw, R. M. (1985) 'Palaeogeomorphology and the occurrence of diamondiferous placer deposits in Koidu, Sierra Leone', *Journal of the Geological Society of London* 142: 789–802.

Thomson, A. L. (ed.) (1964) *A New Dictionary of Birds*, London: Nelson, for the British Ornithologists' Union.

Thomson, J. A. (1931) 'Biology and human progress', in W. Rose (ed.) *An Outline of Modern Knowledge*, pp. 204–251, London: Victor Gollancz.

Thornthwaite, C. W. (1931) 'The climates of North America', *Geographical Review* 21: 633–654.

Thornthwaite, C. W. (1948) 'An approach toward a rational classification of climate', *Geographical Review* 38: 55–94.

Thornton, I. W. B. and New, T. R. (1988) 'Krakatau invertebrates: the 1980s fauna in the context of a century of recolonization', *Philosophical Transactions of the Royal Society of London* 322B: 493–522.

Thorpe, R. S. and Brown, R. P. (1989) 'Microgeographic variation in the colour pattern of the lizard *Gallotia galloti* within the island of Tenerife: distribution, pattern and hypothesis testing', *Biological Journal of the Linnean Society* 38: 303–322.

Tilman, D. and Wedin, D. (1991) 'Oscillations and chaos in the dynamics of a perennial grass', *Nature* 353: 653–655.

Timpson, M. E., Richardson, J. L., Keller, L. P., and McCarthy, G. J. (1986) 'Evaporite mineralogy associated with saline seeps in southwestern North Dakota', *Soil Science Society of America Journal* 50: 490–493.

Tivy, J. (1982) *Biogeography: A Study of Plants in the Ecosphere*, 2nd edn, London and New York: Longman.

Toon, O. B., Pollack, J. B., Ackerman, T. P., Turco, R. P., McKay, C. P., and Liu, M. S. (1982) 'Evolution of an impact-generated dust cloud and its effects on the atmosphere', in L. T. Silver and P. H. Schultz (eds) *Geological Implications of Impacts of Large Asteroids and Comets on the Earth*, pp. 187–200, Geological Society of America Special Paper 190.

Tranquillini, W. (1979) *Physiological Ecology of the Alpine Timberline: Tree Existence at High Altitudes with Special Reference to the European Alps* (Ecological Studies, vol. 31), Berlin: Springer.

Troeh, F. R. (1964) 'Landform parameters correlated to soil drainage', *Soil Science Society of America Proceedings* 28: 808–812.

Troll, C. (1939) 'Luftbildplan und ökologische Bodenforschung', *Zeitschrift der Gesellschaft für Erdkunde zu Berlin* 1939: 241–298.

Troll, C. (1971) 'Landscape ecology (geoecology) and biogeocoenology – a terminological study', *Geoforum* 8: 43–46.

Troll, C. (1972) 'Geoecology and the world-wide differentiation of high-mountain ecosystems', in C. Troll (ed.) *Geoecology of the High-Mountain Regions of Eurasia*, pp. 1–16, Wiesbaden: Franz Steiner.

Turner, J. R. G. and Lennon, J. J. (1989) Reply under 'Species richness and the energy theory', *Nature* 340: 351.

Turner, J. R. G., Gatehouse, C. M., and Corey, C. A. (1987) 'Does solar energy control organic diversity? Butterflies, moths and the British climate', *Oikos* 48: 195–205.

Turner, J. R. G., Lennon, J. J., and Lawrenson, J. A. (1988) 'British bird species distributions and the energy theory', *Nature* 335: 539–541.

Turner, M. G. (ed.) (1987) *Landscape Heterogeneity and Disturbance* (Ecological Studies, vol. 64), New York: Springer.

Turner, M. G. (1989) 'Landscape ecology: the effect of pattern on process', *Annual Review of Ecology and Systematics* 20: 171–197.

Turner, M. G. and Dale, V. H. (1991) 'Modeling landscape disturbance', in M. G. Turner and R. H. Gardner (eds) *Quantitative Methods in Landscape Ecology: the Analysis and Interpretation of Landscape Heterogeneity* (Ecological Studies, vol. 82), pp. 323–351, New York: Springer.

Turner, M. G. and Gardner, R. H. (1991) 'Quantitative methods in landscape ecology: an introduction', in M. G. Turner and R. H. Gardner (eds) *Quantitative Methods in Landscape Ecology: the Analysis and Interpretation of Landscape Heterogeneity* (Ecological Studies, vol. 82), pp. 3–14, New York: Springer.

Turner, M. G. and Stratton, S. P. (1987) 'Fire, grazing, and the landscape heterogeneity of a Georgia barrier island', in M. G. Turner (ed.) *Landscape Heterogeneity and Disturbance* (Ecological Studies, vol. 64), pp. 85–101, New York: Springer.

Turner, M. G., Dale, V. H., and Gardner, R. H. (1989) 'Predicting across scales: theory development and testing', *Landscape Ecology* 3: 245–252.

Turner, M. G., Romme, W. H., Gardner, R. H., O'Neill, R. V., and Kratz, T. K. (1993) 'A revised concept of landscape equilibrium: disturbance and stability on scaled landscapes', *Landscape Ecology* 8: 213–227.

Tüxen, R. (1931/32) 'Die Pflanzensoziologie in ihren Beziehungen zu den Nachbarwissenschaften', *Der Biologe* 8: 180–187.

Udvardy, M. D. F. (1982) 'A biogeographical classification system for terrestrial environments', in J. A. McNeely and K. R. Miller (eds) *National Parks, Conservation, and Development: the Role of Protected Areas in Sustained Society*, pp. 34–38, Washington, DC: Smithsonian Institution Press.

Urey, H. C. (1973) 'Cometary collisions and geological periods', *Nature* 242: 32–33.

Van Cleve, K., Chapin III, F. S., Dyrness, C. T., and Viereck, L. A. (1991) 'Element cycling in taiga forests: state-factor control', *BioScience* 41: 78–88.

Van Cleve, K. and Yarie, J. (1986) 'Interaction of temperature, moisture, and soil chemistry in controlling nutrient cycling and ecosystem development in the taiga of Alaska', in K. Van Cleve, F. S., Chapin III, P. W. Flanagan, L. A. Viereck, and C. T. Dyrness (eds) *Forest Ecosystems in the Alaskan Taiga: A Synthesis of Structure and Function* (Ecological Studies, vol. 57), pp. 160–189, New York: Springer.

Van Es, H. M., Cassel, D. K., and Daniels, R. B. (1991) 'Infiltration variability and correlations with surface soil properties for an eroded Hapludult', *Soil Science Society of America Journal* 55: 486–492.

Van Valen, L. (1973) 'A new evolutionary law', *Evolutionary Theory* 1: 1–30.

van Wilgen, B. W., Richardson, D. M., Kruger, F. J., and van Hensbergen, H. J. (eds) (1992) *Fire in South African Mountain Fynbos: Ecosystem, Community and Species Response at Swartboskloof* (Ecological Studies, vol. 93), New York: Springer.

Vaughan, T. A. (1954) 'Mammals of the San Gabriel Mountains of California, *University of Kansas Publications, Museum of Natural History* 7: 513–582.

Vaughan, T. A. (1978) *Mammalogy*, 2nd edn, Philadelphia: W. B. Saunders.

Veblen, T. T., Ashton, D. H., Schlegel, F. M., and Veblen, A. T. (1977) 'Plant succession in a timberline depressed by vulcanism in south-central Chile', *Journal of Biogeography* 4: 275–294.

Veblen, T. T., Hadley, K. S., Nel, E. M., Kitzberger, T., Reid, M., and Villalba, R. (1994) 'Disturbance regime and disturbance interactions in a Rocky Mountain subalpine forest', *Journal of Ecology* 82: 125–135.

Venables, L. S. V. and Venables, U. M. (1955) *Birds and Mammals of Shetland*, Edinburgh: Oliver & Boyd.

Vernadsky, V. I. (1926) *Biosfera*, Leningrad: Nauchoe Khimikoteknicheskoe Izdatelstvo.

Vernadsky, V. I. (1929) *La Biosphère*, Paris: Félix Alcan.

Vernadsky, V. I. (1944) 'Problems of biogeochemistry, II. The fundamental matter-energy difference between the living and the inert bodies of the biosphere', translated by George Vernadsky, edited and condensed by G. E. Hutchinson, *Transactions of the Connecticut Academy of Arts and Sciences* 35: 483–517.

Vernadsky, V. I. (1945) 'The biosphere and the noösphere', *American Scientist* 33: 1–12.

Vernadsky, V. I. (1986) *The Biosphere*, translated, abridged edn of *La Biosphère*, London: Synergetic Press.

Viles, H. A. (ed.) (1988) *Biogeomorphology*, Oxford: Basil Blackwell.

Vink, A. P. A. (1983) *Landscape Ecology and Land Use*, translated by the author and edited by D. A. Davidson, London and New York: Longman.

Volterra, V. (1926) 'Fluctuations in the abundance of species considered mathematically', *Nature* 188: 558–560.

Volterra, V. (1931) *Leçons sur la Théorie mathématique de la Lutte pour la Vie*, Paris: Gauthier-Villars.

Vreeken, W. J. (1973) 'Soil variability in small loess watersheds: clay and organic matter content', *Catena* 2: 321–336.

Walker, D. A. and Everett, K. R. (1991) 'Loess ecosystems of northern Alaska: regional gradient and toposequence at Prudhoe Bay', *Ecological Monographs* 61: 437–464.

Walker, D. A., Binnian, E., Evans, B. M., Lederer, N. D., Nordstrand, E., and Webber, P. J. (1989) 'Terrain, vegetation and landscape evolution of the R4D research site, Brooks Range Foothills, Alaska', *Holarctic Ecology* 12: 238–261.

Walker, E. P. (1968) *Mammals of the World*, 2nd edn, Baltimore: Johns Hopkins University Press.

Walker, T. W. and Adams, A. F. R. (1959) 'Studies on soil organic matter: 2. Influence of increased leaching at various stages of weathering on levels of carbon, nitrogen, sulfur, and organic and total phosphorus', *Soil Science* 87: 1–10.

Wallace, A. R. (1869) *The Malay Archipelago: the Land of the Orang-utan, and the Bird of Paradise; A Narrative of Travel with Studies of Man and Nature*, 2 vols, London: Macmillan & Co.

Wallace, A. R. (1880) *Island Life: Or, the Phenomena and Causes of Insular Faunas and Floras, Including a Revision and Attempted Solution of the Problem of Geological Climates*, London: Macmillan & Co.

Walter, H. (1962) *Die Vegetation der Erde in ökophysiologischer Betrachtung. I. Die tropischen und subtropischen Zonen*, Stuttgart and Jena: Gustav Fischer.

Walter, H. (1973) *Vegetation of the Earth in Relation to Climate and Eco-physiological Conditions*, translated by J. Weiser, London: English Universities Press.

Walter, H. (1985) *Vegetation of the Earth and Ecological Systems of the Geo-Biosphere*, 3rd revised and enlarged edn, translated from the 5th revised German edn by O. Muise, Berlin: Springer.

Walter, H. and Breckle, S.-W. (1985) *Ecological Systems of the Geobiosphere. Vol. 1. Ecological Principles in Global Perspective*, translated by S. Gruber, Berlin: Springer.

Walter, H. and Lieth, H. (1960–1967) *Klimadiagramm-Weltatlas*, Jena: Gustav Fischer.

Wang, K., Chatterton, B. D. E., Attrep, M., Jr, and Orth, C. J. (1992) 'Iridium abundance maxima at the latest Ordovician mass extinction horizon, Yangtze Basin, China: terrestrial or extraterrestrial?', *Geology* 20: 39–42.

Wang, K., Orth, C. J., Attrep, M., Jr, Chatterton, B. D. E., Hou, H.-F., and Geldsetzer, H. H. J. (1991) 'Geochemical evidence for a catastrophic biotic event at the Frasnian/Famennian boundary in south China', *Geology* 19: 776–779.

Wardle, P. (1974) 'Alpine timberlines', in J. D. Ives and R. G. Barry (eds) *Arctic and Alpine Environments*, pp. 370–402, London: Methuen.

Warren, A. and Cowie, J. (1976) 'The use of soil maps in education, research and planning', in D. A. Davidson (ed.) *Soil Survey Interpretation and Use*, pp. 1–14, Welsh Soils Discussion Group Report No. 17.

Watson, A. J. and Lovelock, J. E. (1983) 'Biological homeostasis of the global environment: the parable of Daisyworld', *Tellus* 35: 284–289.

Watt, A. S. (1924) 'On the ecology of British beechwoods with special reference to their regeneration. II. The development and structure of beech communities on the Sussex Downs', *Journal of Ecology* 12: 145–204.

Watt, A. S. (1947) 'Pattern and process in the plant community', *Journal of Ecology* 35: 1–22.

West, L. T., Wilding, L. P., Stahnke, C. R., and Hallmark, C. T. (1988) 'Calciustolls in central Texas: I. Parent material uniformity and hillslope effects on carbonate enriched horizons', *Journal of the Soil Science Society of America* 52: 1722–1731.

Westbroek, P. (1991) *Life as A Geological Force: Dynamics of the Earth*, New York and London: W. W. Norton.

White, G. (1789) *The Natural History of Selbourne with Observations on Various Parts of Nature, and the Naturalist's Calendar*, London: B. White & Son.

White, P. S. and Pickett, S. T. A. (1985) 'Natural disturbance and patch dynamics: an introduction', in S. T. A. Pickett and P. S. White (eds) *The Ecology of Natural Disturbance and Patch Dynamics*, pp. 3–13, Orlando, Florida: Academic Press.

Whitehead, P. J., Bowman, D. M. J. S., and Tidemann, S. C. (1992) 'Biogeographic patterns, environmental correlates and conservation of avifauna in the Northern Territory, Australia', *Journal of Biogeography* 19: 151–161.

Whitfield, W. A. D. and Furley, P. A. (1971) 'The relationship between soil patterns and slope form in the Ettrick Association, south-east Scotland', in D. Brunsden (ed.) *Slopes: Form and Process* (Institute of British Geographers, Special Publication No. 3), pp. 165–175, Kensington Gore, London: Institute of British Geographers.

Whitmore, T. C. (1975) *Tropical Rain Forests of the Far East*, Oxford: Oxford University Press.

Whittaker, R. H. (1954) 'The ecology of serpentine soils. IV. The vegetational response to serpentine soils', *Ecology* 35: 275–288.

Whittaker, R. H. (1967) 'Gradient analysis of vegetation', *Biological Reviews* 42: 207–264.

Whittaker, R. H. (ed.) (1978) *Ordination of Plant Communities*, The Hague: W. Junk.

Whittaker, R. J. (1987) 'An application of detrended correspondence analysis and non-metric multidimensional scaling to the identification and analysis of environmental factor complexes and vegetation structures', *Journal of Ecology* 75: 363–376.

Whittaker, R. J. (1989) 'The vegetation of the Storbreen gletschervorfeld, Jotunheimen, Norway. III. Vegetation–environment relationships', *Journal of Biogeography* 18: 41–52.

Whittlesey, D. (1954) 'The regional concept and the regional method', in P. E. James and C. F. Jones (eds) *American Geography: Inventory and Prospect*, pp. 19–68, Syracuse, New York: Syracuse University Press.

Wilde, S. A. (1946) *Forest Soils and Forest Growth*, Waltham, Massachusetts: Chronica Botanica.

Williamson, M. (1981) *Island Populations*, Oxford: Oxford University Press.

Wolbach, W. S., Lewis, R. S, and Anders, E. (1985) 'Cretaceous extinctions: evidence for wildfires and search for meteoritic material', *Science* 230: 167–170.

Wolbach, W. S., Gilmour, I., Anders, E., Orth, C., and Brooks, R. R. (1988) 'Global fire at the Cretaceous–Tertiary boundary', *Nature* 334: 665–669.

Woodward, F. I. (1987) *Climate and Plant Distribution*, Cambridge: Cambridge University Press.

Worster, D. (1990) 'The ecology of order and chaos', *Environmental History Review* 14: 1–18.

Wright, D. H. (1983) 'Species–energy theory: an extension of species–area theory', *Oikos* 41: 496–506.

Wylie, J. L. and Currie, D. J. (1993a) 'Species–energy theory and patterns of species richness: I. Patterns of bird, angiosperm, and mammal species richness on islands', *Biological Conservation* 63: 137–144.

Wylie, J. L. and Currie, D. J. (1993b) 'Species–energy theory and patterns of species richness: II. Predicting mammal species richness on isolated nature reserves', *Biological Conservation* 63: 145–148.

Yaalon, D. H. (1965) 'Downward movement and distribution of anions in soil profiles with limited wetting', in E. G. Hallsworth and D. V. Crawford (eds) *Experimental Pedology*, pp. 157–164, London: Butterworths.

Yaalon, D. H. (1971) 'Soil-forming processes in time and space', in D. H. Yaalon (ed.) *Paleopedology – Origin, Nature and Dating of Paleosols*, pp. 29–39, Jerusalem, Israel: International Society of Soil Science.

Yaalon, D. H. (1975) 'Conceptual models in pedogenesis: can soil-forming functions be solved?', *Geoderma* 14: 189–205.

Yaalon, D. H., Brenner, I., and Koyumdjiski, H. (1974) 'Weathering and mobility sequence of minor elements on a basaltic pedomorphic surface', *Geoderma* 12: 233–244.

Zimmerman, E. A. W. von (1777) *Specimen Zoologicae Geographicae, Quadrupedum Domicilia et Migrationes Sistens. Dedit Tabulamque Mundi Zoographicam Adjunxit Eberh Aug Guilielm Zimmerman*, Leyden: T. Haak.

INDEX

Note: Italicized numbers refer to pages on which authors or subjects are cited only in figures or tables.